# Qualitative and Computational Aspects of Dynamical Systems

*Edited by Kamal Shah, Bruno Carpentieri and Arshad Ali*

Published in London, United Kingdom

Qualitative and Computational Aspects of Dynamical Systems
http://dx.doi.org/10.5772/intechopen.100719
Edited by Kamal Shah, Bruno Carpentieri and Arshad Ali

Assistant to the Editors: Eiman Ijaz

Contributors
Sanae Elouahdani, Hamid Boukhal, El Mahjoub Chakir, Ahmed Gaga, Houda Elyaakoubi, Mustapha Makhloul, Abdelaziz Ahmed, Abdessamad Didi, Mohamed Bencheikh, Carlos Rodríguez Lucatero, Borut Žalik, David Podgorelec, Niko Lukač, Krista Rizman Žalik, Domen Mongus, Praveen Kumar Sharma, Shivram Sharma, Jitendra Kaushik, Palash Goyal, Taj Munir, Eiman Ijaz, Johar Ali, Abbas Khan, Muhammad Shafiq, Alexandru G. Gabriel Gheorghe, Mihai E. Marin, Atimad Harir, Said Melliani, Lalla Saadia Chadli, Victor Akinsola



First published in London, United Kingdom, 2023 by IntechOpen
IntechOpen is the global imprint of INTECHOPEN LIMITED, registered in England and Wales, registration number: 11086078, 5 Princes Gate Court, London, SW7 2QJ, United Kingdom

British Library Cataloguing-in-Publication Data
A catalogue record for this book is available from the British Library

Additional hard and PDF copies can be obtained from orders@intechopen.com

Qualitative and Computational Aspects of Dynamical Systems
Edited by Kamal Shah, Bruno Carpentieri and Arshad Ali
p. cm.
Print ISBN 978-1-80356-566-8
Online ISBN 978-1-80356-567-5
eBook (PDF) ISBN 978-1-80356-568-2

# Meet the editors

Dr. Kamal Shah is a senior researcher in the Department of Mathematics and Sciences at Prince Sultan University, Riyadh, Saudi Arabia. His previous appointment was as an associate professor in the Department of Mathematics at the University of Malakand, Pakistan. His research interests are fractional calculus, nonlinear analysis, and numerical solutions of differential equations. He has published hundreds of research articles in internationally recognized journals and serves as an academic editor for many journals of high repute. He has supervised several doctoral and master's students in the field of applied mathematics.

Bruno Carpentieri obtained a Laurea degree in applied mathematics from Bari University in 1997, and a Ph.D. in computer science from the Institut National Polytechnique de Toulouse (INPT), France. After holding various post-doctoral positions, he became an assistant professor at the Bernoulli Institute for Mathematics, Computer Science and Artificial Intelligence, University of Groningen, and then a Reader in applied mathematics at Nottingham Trent University, UK. Since May 2017, he has been an associate professor in applied mathematics at the Faculty of Computer Science, Free University of Bozen-Bolzano. His research interests are in the fields of applied mathematics, numerical linear algebra, and high-performance computing. Bruno Carpentieri has served as a member of several scientific advisory boards in computational mathematics. He is an editorial board member of the *Journal of Applied Mathematics*, an editorial committee member of *Mathematical Reviews* (American Mathematical Society), and a reviewer of about 30 numerical analysis journals. He has co-authored about 50 publications in peer-reviewed scientific journals.

Dr. Arshad Ali comes from the beautiful valley of Swat, Khyber Pakhtunkhwa, Pakistan and is currently teaching in a government school in Swat. He obtained his master's degree in mathematics from the Government Post-graduate Jehan Zeb College Swat, Khyber Pakhtunkhwa, in 2009, and his MPhil in 2017. His Ph.D. in applied mathematics was supervised by Dr. Kamal Shah. He has published several research articles in highly reputed journals.

# Contents

# Preface

Dynamical systems play an essential role in describing various real-world processes and phenomena that evolve over time. In the analysis of dynamical systems, several important theoretical and computational aspects need to be addressed, from the basic and more advanced theory to the development of efficient numerical methods, and finally their application to the simulation of practical real-world problems. The book discusses these ideas, covering basic techniques, including some review material, and more advanced topics such as the fuzzy concept and the application of discrete calculus to the study of specific problems.

The book contains nine chapters. Chapter 1 reviews the Laplace transform and its applications in the study of solutions of dynamical systems. In Chapter 2, numerical methods based on variants of the fourth-order Runge–Kutta and Euler algorithms are discussed. Chapter 3 is devoted to the development and analysis of an efficient region-merging algorithm in raster space. Chapter 4 investigates the application of discrete mathematics for programming discrete mathematics calculations. Chapter 5 presents a criticality study of fast critical experimental benchmarks using MCNP code to qualify different evaluations. Chapters 6 and 7 are about the use of fuzzy concepts to investigate problems of dynamical systems. In Chapter 8, a case study of dynamical systems arising in circuit analysis is proposed. Finally, in Chapter 9, a dynamical system of infectious disease is studied numerically using a non-standard finite difference method.

**Kamal Shah**
Department of Mathematics and Sciences,
Prince Sultan University,
Riyadh, Saudi Arabia

**Bruno Carpentieri**
Faculty of Computer Science,
Free University of Bozen-Bolzano,
Bolzano, Italy

**Arshad Ali**
Department of Mathematics
University of Malakand,
Chakdara Dir(L), KPK, Pakistan

Chapter 1

# A Review Note on Laplace Transform and Its Applications in Dynamical Systems

*Shivram Sharma, Praveen Kumar Sharma and Jitendra Kaushik*

## Abstract

Laplace Transform is one of the essential transform techniques. It has many applications in engineering and science. The Laplace transform techniques can be used to solve various partial differential equations and ordinary differential equations that cannot be resolved using conventional techniques. The Laplace transform approach is practically the essential functional method for engineers. The Laplace transform and variations like the fuzzy Laplace transform are advantageous because they directly solve issues such as initial value problems, fuzzy initial value problems, and non-homogeneous differential equations without first resolving the corresponding homogeneous equation. This chapter uses the Laplace transform and its variations to dynamical systems.

**Keywords:** Laplace transform, Inverse Laplace transform, Fuzzy Laplace transform properties, applications, initial value problem, electrical circuits

## 1. Introduction

A problem can be solved easily in another domain using an integral transform, and the solution is then transformed using an inverse transform back to the original domain. The Laplace transformation is one such transformation that Pierre-Simon Laplace found in 1785. A popular integral transform in mathematics with several uses in science and engineering is the Laplace Transform. A transformation from the time domain, where the inputs and outputs are time functions, to the frequency domain, where the inputs and outputs are functions of complex angular frequency, is what the Laplace Transform entails. The Laplace transform is frequently used to convert a differential equation system into an algebraic equation system and to multiply a convolution [1–5]. It entails decomposing a system of differential equations into a set of linear equations that must be solved and then applying the inverse Laplace transform to return the answer to the time domain. In many circumstances, the result is divided into "patterns," for which the inverse transform is known. The inverse Laplace transform is then used to return the solution to the time domain.

**Definition 1.1.** Suppose that $f$ is real or complex-valued function of time $t > 0$, and $p$ is a real or complex parameter, then the Laplace transform of $f(t)$ is defined as

$$\mathcal{L}[f(t)] = \int_0^\infty e^{-pt} f(t)\, dt = F(p) \tag{1}$$

(Provided the integral defined in (1) exists)

**Example 1.1.** If $f(t) = 1$ for $t \geq 0$, then The Laplace Transform of this function $f(t) = 1$ can be obtained by using (1) as:

$$\mathcal{L}\,[1] = \int_0^\infty e^{-pt}.1\, dt = \int_0^\infty e^{-pt}\, dt = \lim_{t\to\infty}\left[\frac{e^{-pt}}{-p}\right]_0^t = \frac{1}{p}$$

Thus, we have Laplace transform of $f(t) = 1$, which is given by $\mathcal{L}\,[1] = \frac{1}{p}$.

Taking this formula's, Inverse Laplace Transform ($\mathcal{L}^{-1}$), we get the inverse Laplace Transform of $\frac{1}{p}$, which is given by $\mathcal{L}^{-1}\left(\frac{1}{p}\right) = 1$.

Similarly, following the above procedure, we can find the Laplace transform and inverse Laplace transforms of other valuable functions.

Some essential formulae of LT and ILT are given below in tabular form **Table 1**, [5], which are very useful for finding out the results of problems/systems by the Laplace transform method.

| S. No | Laplace transform | Inverse Laplace transform |
|---|---|---|
| 1 | 1 | $\frac{1}{p}$ |
| 2 | $e^{at}$ | $\frac{1}{p-a}$ |
| 3 | $t^n$, $n = 1, 2, 3$ | $\frac{n!}{p^{n+1}}$ |
| 4 | $t^n, n > -1$ | $\frac{\Gamma(n+1)}{p^{n+1}}$ |
| 5 | $\sin(at)$ | $\frac{a}{p^2+a^2}$ |
| 6 | $\cos(at)$ | $\frac{p}{p^2+a^2}$ |
| 7 | $\sinh(at)$ | $\frac{a}{p^2-a^2}$ |
| 8 | $\cosh(at)$ | $\frac{s}{p^2-a^2}$ |
| 9 | $f(ct)$ | $\frac{1}{c}F\left(\frac{p}{c}\right)$ |
| 10 | $e^{ct}f(t)$ | $F(\mathrm{p}-c)$ |
| 11 | $t^n, f(t)$, $n = 1, 2, 3 \ldots$ | $(-1)^n F^{(n)}(p)$ |
| 12 | $\frac{1}{t}f(t)$ | $\int_p^\infty F(v)dv$ |
| 13 | $\int_0^t f(v)dv$ | $\frac{F(p)}{p}$ |
| 14 | $\int_0^t f(t-\Gamma)g(\Gamma)d\Gamma$ | $F(p)G(p)$ |
| 15 | $f(t+T) = f(t)$ | $\frac{\int_0^T e^{pt} f(t)dt}{1-e^{pT}}$ |
| 16 | $f'(t)$ | $pF(p - f(0)$ |
| 17 | $f''(t)$ | $p^2F(p) - pf(0) - f'(0)$ |

**Table 1.**
*Formulae of LT and ILT.*

## 2. A sufficient condition for the Existence of Laplace transforms any function

The existence of LT of any function$f(t)$ depends on the piecewise continuity of the function on the interval $[0\infty)$ and the exponential order of the function.

If a function is piecewise continuous on the interval $[0, \infty)$ and has exponential order $\alpha$ (for $t \geq 0$, for some $p > \alpha$), then the LT of function $f(t)$ exists.

The above conditions are only sufficient conditions but not necessary conditions for the existence of the Laplace Transform of any function. A function may have Laplace transform even if it violates the above conditions/existence conditions.

For example: The function $f(t) = t^{-\frac{1}{2}}$ is not continuous at $t = 0$, even though its Laplace transform exists.

Moreover, when the analysis depends on a mathematical model of differential equations due to the uncertainty of future aspects, we require mathematical tools to prescience other risks. We can categorize uncertainties into two categories. 1. Possible sets or fuzzy sets handle uncertainty. 2. Unpredictable uncertainty, which probability models shall deal with.

The publication of a seminal paper by Zadeh [6], a computer scientist from the University of California, was a crucial turning point in developing the modern concept of uncertainty. In his seminal paper, the USA was the first to introduce the concept of fuzzy set theory as a new way to represent vagueness in our daily lives. Zadeh defined "fuzzy sets" in his theory as sets with ill-defined bounds. This idea is employed and found to be superior for solving issues across many fields.

Recently, the analysis of dynamical systems has obtained more attention due to mathematical models of fuzzy applications through Laplace transforms. Hence, pre-eminence and optimal solution are required through fuzzy concepts of H- derivatives and SGH derivatives (1) gH-derivative, g-derivative and gr-derivative. Few researchers such as Najariyan, & Farahi [7], Najariyan, & Farahi [8], Najariyan, & Zhao [9], Mazandarani & Pariz [10] and Najariyan & Zhao [11] attempt the work of optimal solution through the dynamic systems of the fuzzy approach.

Allahviranloo and Ahmadi [12] recently proposed a fuzzy Laplace transformation for first-order fuzzy differentiation equations. They keep focused on the fuzzy Laplace transform concept yet to explain the fuzzy valued conditions. Then, Salahshour and Allahviranloo [13] focused on first- and second-order derivative Laplace transform for linear, continuous, uniform, and convergence problems. They considered FIVP, H-differentiation, and second-order derivatives. The large-size fuzzy-valued function can be executed by fuzzy Laplace transformation.

Using fuzzy differential equations (FDEs) to model dynamic systems with uncertainty makes sense. First-order linear fuzzy differential equations are among the most fundamental FDEs in several applications. Chang and Zadeh initially presented the fuzzy derivative concept in 1972 [14]. Kandel and Byatt [15, 16] analyzed fuzzy dynamical issues using the idea of FDEs. The fuzzy Laplace transform method solves FDEs and their fuzzy beginning and boundary value problems. Fuzzy Laplace transforms reduce an FDE to an algebraic problem, which facilitates its solution.

Here in this chapter, we aim to study the applications of Laplace transform in dynamical systems, so we present our study under four subsections 2.1, 2.2, 2.3, and 2.4. At the end of this chapter, we give a brief conclusion about the proposed research.

# 3. Main results/discussion/ application of Laplace transform in a dynamic system

In this chapter, our main aim is to apply Laplace transform in a dynamic system.

**A dynamical system** [17] is one in which something evolves over time or in which a function describes the time dependence of a point in the surrounding space. For instance, mathematical representations of the pendulum of a clock, the flow of water through a pipe, the number of fish in a lake each spring, population growth, and so forth.

Both a continuous timeline and discrete time increments can be used to depict a dynamic system.

**Discrete-time dynamical system [17]**

$$x_t = F(x_{t-1}, t) \tag{2}$$

This type of model is called a difference equation, a recurrence equation, or an iterative map (if the right-hand side is not dependent on)

**Continuous-time dynamical system [17]**

$$\frac{dx}{dt} = F(x, t) \tag{3}$$

This type of model is called a differential equation.

The system's state variable at time t in both scenarios is $x_t$ or $x$, which may take a scalar or vector value. The rule by which the system changes its state over time is determined by a function called F.

Differential equations often model dynamical systems.

So, here we discuss an application of Laplace transform and its variant (e.g., fuzzy Laplace transform) to solve differential equations /electrical circuits/mechanical systems/ fuzzy differential equations:

## 3.1 Solution of differential equations (including IVP and system of simultaneous differential equations) by using Laplace transform technique

**Example 2.1.1.** Consider an IVP

$$\frac{dy}{dt} + y = sint; y(0) = 1$$

Taking the Laplace transform of both sides of the above equation, we have

$$\mathcal{L}\left[\frac{dy}{dt}\right] + \mathcal{L}[y] = \mathcal{L}[sint]$$

$$p\mathcal{L}[y(t)] - y(0) + \mathcal{L}[y(t)] = \frac{1}{p^2 + 1}$$

$$(p + 1)\mathcal{L}[y(t)] = 1 + \frac{1}{p^2 + 1}$$

$$\mathcal{L}[y(t)] = \frac{1}{p + 1} + \frac{1}{(p^2 + 1)(p + 1)}$$

Taking the Inverse Laplace transform of both sides of the above equation, we have

$$y(t) = \mathcal{L}^{-1}\left[\frac{1}{p+1}\right] + \mathcal{L}^{-1}\left[\frac{1}{(p^2+1)(p+1)}\right]$$

$$y(t) = e^{-t} + \frac{1}{2}\mathcal{L}^{-1}\left[\frac{1}{(p+1)} - \frac{p}{p^2+1} + \frac{1}{p^2+1}\right]$$

$$y(t) = e^{-t} + \frac{1}{2}[e^{-t} - \cos t + \sin t\,]$$

$$y(t) = \frac{3}{2}e^{-t} + \frac{1}{2}[\,\sin t - \cos t]$$

Which is the required solution for given IVP.

**Remark 2.1.1.** Mathematical modeling of the system is a must for analyzing/predicting the nature and behavior of any physical system. There are many physical systems wherein we get IVP, for example, the growth and decay population model of Malthus, which is represented by $\frac{dy}{dt} = ky$ where $y(t)$ is the population at any time $t$ and the constant of proportionality k is the growth constant and is used for finding the bacteria in a culture, life of radioactive substance and newtons law of cooling. A solution to these problems can be obtained by following the procedure in example 2.1.1 using Laplace transform techniques.

**Example 2.1.2.** Consider a system of simultaneous equations

$$\begin{cases} \dfrac{dx}{dt} + y = \sin t \\ \dfrac{dy}{dt} + x = 1 + \cos t \end{cases} ;with\ x(0) = 0\&y(0) = 0 \tag{4}$$

Using Laplace transform, we have

$$\mathcal{L}\left[\frac{dx}{dt} + y\right] = \mathcal{L}[\,\sin t],\mathcal{L}\left[\frac{dy}{dt} + x\right] = \mathcal{L}[t + \cos t]$$

$$p\ \mathcal{L}[x(t)] - x(0) + \mathcal{L}[y(t)] = \frac{1}{p^2+1},\mathcal{L}[y(t)] - y(0) + \mathcal{L}[x(t)] = \frac{1}{p} + \frac{p}{p^2+1}$$

$$p\ \mathcal{L}[x(t)] + \mathcal{L}[y(t)] = \frac{1}{p^2+1},\mathcal{L}[y(t)] + \mathcal{L}[x(t)] = \frac{1}{p} + \frac{p}{p^2+1}$$

On solving the above pair of equations for $\mathcal{L}[(et)]$, we have

$$\mathcal{L}[y(t)] = \frac{2}{p-1} - \frac{2}{(p-1)(p^2+1)}$$

Taking the Inverse Laplace Transform of both sides of the above equation, we have

$$y(t) = \mathcal{L}^{-1}\left[\frac{2}{p-1}\right] - \mathcal{L}^{-1}\left[\frac{2}{(p-1)(p^2+1)}\right]$$

$$y(t) = 2e^{t} - \mathcal{L}^{-1}\left[\frac{1}{(p-1)} - \frac{p}{p^2+1} - \frac{1}{p^2+1}\right]$$

$$y(t) = e^t - \cos t - \sin t \tag{5}$$

Putting the above value of $y(t)$ in the second equation of equation no. (4)

$$x(t) = 1 + \cos t - \frac{d}{dt}[e^t - \cos t - \sin t]$$

$$x(t) = 1 + \cos t - [e^t + \sin t - \cos t]$$

$$x(t) = 1 - e^t - \sin t \tag{6}$$

Equations 2nd and 3rd together give the solution of the given system of simultaneous differential equations.

**Remark 2.1.2.** In many physical problems/situations like particle movement along any curve at time t, two tanks in mixing problems, and two circuits in electrical networks, etc., we get a system of ordinary differential equations. The solution to all these problems can also be obtained by following the procedure in example 2.1.2 using Laplace transform techniques.

## 3.2 Solution/response/current in electrical circuits by using Laplace transform technique

This section finds the electrical circuits' response/solution/current $i(t)$ using Laplace transform techniques. Many authors have done work in this field, but their results were different and proved using a different methodology. Some basic rules and principles of the area are required to prove results under this section. Although there are many advanced books where we can get these basic rules/ definitions /principles, we refer to [5] for this purpose.

**Definition 2.2.1 [5]. Kirchhoff's Laws**

I. **Voltage Law:** the algebraic sum of the voltage drops around any closed circuit equals the circuit's resultant electromotive force (EMF).

II. **Current Law:** at a junction or node, current coming is current going.

**Example 2.2.1. (Response/Solution/Current $i(t)$ in the L-R Series Circuit)**

A simple R-L series circuit is shown in **Figure 1**, where resistance (R) and Inductance (L) are in series with voltage source E(t). Let $i(t)$ be the current flowing in the circuit at any time t.

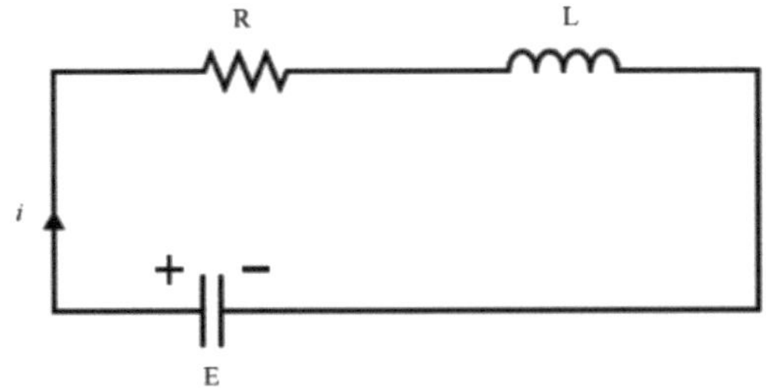


**Figure 1.**
*R-L series circuit.*

Then, the model of the L-R series circuit is given by:
$R\,i + L\frac{di}{dt} = E(t)$ (By voltage law)

$$\frac{di}{dt} + \frac{R}{L}i = \frac{E}{L} \tag{7}$$

It is assumed that the current is initially zero, i.e., $i = 0$ *at* $t = 0$.
Taking the Laplace transform of both sides of the above equation, we get

$$\mathcal{L}\left[\frac{di}{dt} + \frac{R}{L}i\right] = \mathcal{L}\left[\frac{E}{L}\right]$$

$$\mathcal{L}\left[\frac{di}{dt}\right] + \frac{R}{L}\mathcal{L}[i] = \frac{E}{L}\mathcal{L}[1]$$

$$p\;\mathcal{L}[i(t)] - i(0) + \frac{R}{L}\mathcal{L}[i(t)] = \frac{E}{L}\left[\frac{1}{p}\right]$$

$$\mathcal{L}[i(t)] = \frac{E}{L}\left[\frac{1}{p\left(p + \frac{R}{L}\right)}\right] \tag{8}$$

Taking the inverse Laplace transform of both sides of the above equation, we get

$$i(t) = \frac{E}{L}\mathcal{L}^{-1}\left[\frac{1}{p\left(p + \frac{R}{L}\right)}\right]$$

$$i(t) = \frac{E}{R}\mathcal{L}^{-1}\left[\frac{1}{p} - \frac{1}{p + \frac{R}{L}}\right]$$

$$i(t) = \frac{E}{R}\left[1 - e^{-\frac{R}{L}t}\right] \tag{9}$$

Which is the required solution of the given L-R series circuit.

**Example 2.2.2. (Response/Solution/Current $i(t)$ in the RLC circuit)**

In **Figure 2**, an RLC circuit is shown. An RLC circuit is obtained from an RL circuit by adding a capacitor. We have modeled the RL circuit in example 2.2.1, which is given by

$$R\,i + L\frac{di}{dt} = E(t)$$

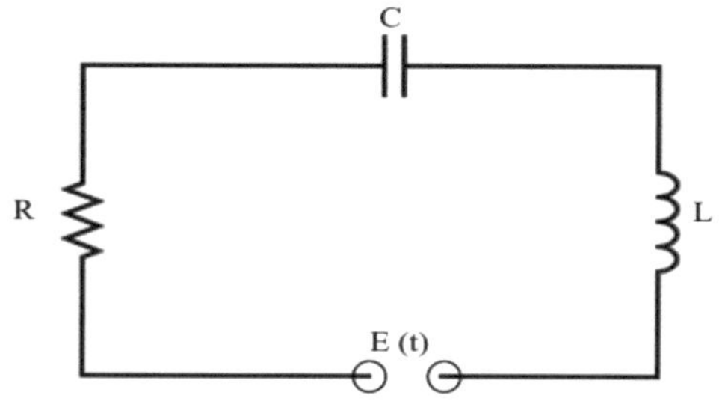


**Figure 2.**
*RLC circuit.*

We obtain the RLC circuit simply by adding the voltage drop Q/C across the capacitor.

Here, current $i(t) = \frac{dQ}{dt}$ (or) $Q(t) = \int i(t)dt$

Assuming a sinusoidal EMF as in the figure. Thus, the model of the RLC circuit is given by:

$$L\frac{di}{dt} + iR + \frac{1}{C}\int i(t)dt = E(t) \tag{10}$$

Differentiating 1st concerning t, we have

$$L\frac{d^2i}{dt^2} + R\frac{di}{dt} + \frac{1}{C}i = E'(t) = E \tag{11}$$

The solution of 2nd will give the current $I$ in the RLC circuit

Taking Laplace transform of both sides of 2nd, we have

$$L\,\mathcal{L}\left[\frac{d^2i}{dt^2}\right] + R\mathcal{L}\left[\frac{di}{dt}\right] + \frac{1}{C}\mathcal{L}[i] = E\mathcal{L}[1]$$

$$(\mathrm{LC})\left[\mathrm{p}^2\mathcal{L}\{\mathrm{i(t)}\} - \mathrm{p\,i(0)} - \mathrm{i'(0)}\right] + (\mathrm{RC})[\mathrm{p}\,\mathcal{L}\{\mathrm{i(t)}\} - \mathrm{i(0)}] + \mathcal{L}[\mathrm{i(t)}] = \mathrm{EC}\left[\frac{1}{\mathrm{p}}\right]$$

$$\left(\mathrm{LCp}^2 + RCp + 1\right)\mathcal{L}[i(t)] = EC\,\frac{1}{\mathrm{p}}$$

(As current and capacitor charge are 0 when t = 0, so $i(0) = 0$, and $i'(0) = 0$)

This implies that

$$\mathcal{L}[\mathrm{i(t)}] = \mathrm{EC}\left[\frac{1}{p(\,\delta\mathrm{p}^2 + \gamma p + 1)}\right] \tag{12}$$

(where $\gamma = RC, \delta = LC$)

Taking Inverse Laplace Transform of both sides of the above equation we have

$$\mathrm{i(t)} = (\mathrm{EC})\mathcal{L}^{-1}\left[\frac{1}{p(1 + \gamma p + \delta p^2)}\right]$$

$$\mathrm{i(t)} = (\mathrm{EC})\mathcal{L}^{-1}\left[\frac{1}{p} - \frac{\gamma + \delta p}{1 + \gamma p + \delta p^2}\right]$$

$$\mathrm{i(t)} = \mathrm{EC} - \mathrm{EC}\,\mathcal{L}^{-1}\left[\frac{(\gamma)}{1 + \gamma p + \delta p^2}\right] - EC\,\mathcal{L}^{-1}\left[\frac{(\delta)p}{1 + \gamma p + \delta p^2}\right]$$

$$\mathrm{i(t)} = \mathrm{EC} - \mathrm{EC}\,(\gamma)\mathcal{L}^{-1}\left[\frac{1}{\left(p + \frac{\gamma}{2\delta}\right)^2 + \left(\frac{1}{\delta} - \frac{\gamma^2}{4\delta^2}\right)}\right] - EC\,(\delta)\mathcal{L}^{-1}\left[\frac{\left(p + \frac{\gamma}{2\delta}\right) - \frac{\gamma}{2\delta}}{\left(p + \frac{\gamma}{2\delta}\right)^2 + \left(\frac{1}{\delta} - \frac{\gamma^2}{4\delta^2}\right)}\right]$$

$$\mathrm{i(t)} = \mathrm{EC} - \mathrm{EC}\,(\gamma)e^{-\frac{\gamma}{2\delta}t}\mathcal{L}^{-1}\left[\frac{1}{p^2 + \left(\frac{1}{\delta} - \frac{\gamma^2}{4\delta^2}\right)}\right] - EC\,(\delta)e^{-\frac{\gamma}{2\delta}t}\mathcal{L}^{-1}\left[\frac{p - \frac{\gamma}{2\delta}}{p^2 + \left(\frac{1}{\delta} - \frac{\gamma^2}{4\delta^2}\right)}\right]$$

This implies that

$$i(t) = EC - EC(\gamma)\left(\frac{2\delta}{\sqrt{4\delta-\gamma^2}}\right)e^{-\frac{\gamma}{2\delta}t}\sin\left(\frac{\sqrt{4\delta-\gamma^2}}{2\delta}\right)t$$
$$-EC(\delta)e^{-\frac{\gamma}{2\delta}t}\left[\cos\left(\frac{\sqrt{4\delta-\gamma^2}}{2\delta}\right)t - \left(\frac{\gamma}{\sqrt{4\delta-\gamma^2}}\right)\sin\left(\frac{\sqrt{4\delta-\gamma^2}}{2\delta}\right)t\right] \quad (13)$$

(where, $\gamma = RC, \delta = LC$)
Which is the required solution of the given RLC circuit.

### 3.3 Solution of mechanical systems by using Laplace transform technique

Some basic rules and principles of the field are required to prove our main result in this section. We refer to [5] for this purpose.

**Example 2.3.1. (Torsional Pendulum Experiment)**

In **Figure 3**, a Torsional Pendulum is demonstrated.
It consists of a disk or rod suspended at the end of the wire.
When the end of the wire is twisted at an angle $\varnothing$
The restoring torque $\tau$ arises
Then by Hooke's Law: $\tau = -k\varnothing$ ... (1) (where $k$ is called the torsional constant).
If the wire is twisted and released, the oscillating system is called a torsional pendulum.
By Newton's Second Law: $\tau = Ia$ ... (2) (where $= \frac{d^2\varnothing}{dt^2}$ )
By 1st and 2nd and we have

$$-k\varnothing = I\frac{d^2\varnothing}{dt^2}$$

Which can be rewritten as

$$\frac{d^2\varnothing}{dt^2} + \frac{k}{I}\varnothing = 0$$

(Or)

$$\frac{d^2\varnothing}{dt^2} + \omega^2\varnothing = 0 \quad (14)$$

This is called the equation of a simple harmonic oscillator, whose angular frequency is $\omega = \sqrt{\frac{k}{I}}$ and period is $T = 2\pi\sqrt{\frac{I}{K}}$

**Figure 3.**
*A Torsional Pendulum.*

Now taking Laplace transform of both sides of the above equation, we have

$$\mathcal{L}\left[\frac{d^2\varnothing}{dt^2}\right] + \omega^2\mathcal{L}[\varnothing] = 0$$

$$[p^2\mathcal{L}\{\varnothing(t)\} - p\varnothing(0) - \varnothing'(0)] + \omega^2\mathcal{L}[\varnothing(t)] = 0$$

[Let, $\varnothing$ (0) = A and$\varnothing'(0) = B$ ]

$$(p^2 + \omega^2)\mathcal{L}[\varnothing(t)] = pA + B$$

$$\mathcal{L}[\varnothing(t)] = \frac{pA + B}{p^2 + \omega^2} \tag{15}$$

Now taking the Inverse Laplace Transform of both sides of Eq. (16)

$$\varnothing(t) = \mathcal{L}^{-1}\left[\frac{pA + B}{p^2 + \omega^2}\right]$$

$$\varnothing(t) = A\,\mathcal{L}^{-1}\left[\frac{p}{p^2 + \omega^2}\right] + B\mathcal{L}^{-1}\left[\frac{1}{p^2 + \omega^2}\right]$$

$$\varnothing(t) = A\cos\omega t + B\frac{\sin\omega t}{\omega} \tag{16}$$

Which is the required solution of the given Mechanical System.

**Remark 2.3.1.** The balance wheel in a clock or wristwatch is an example of a torsional pendulum. All simple harmonic oscillators satisfy a differential equation 3rd of the above example. Hence, the general solution (displacement) of all bodies moving with simple harmonic motion can be obtained by following the procedure given in the above example using Laplace transform technique.

## 3.4 Solution of fuzzy differential equations (FDEs) by using fuzzy Laplace transform

The chapter by Sharma et al. [17] titled "Applications of fuzzy set and fixed point theory in Dynamical systems" and published in the open-access book "Qualitative and Computational Aspects of Dynamical Systems" with the ISBN: 978-1-80356-567-5) contains a detailed discussion of this section (one can refer to this chapter and can go through the definition 2.1, theorem2.2, formulae 2.3, and remark 2.5 of this chapter to understand the concept of this sub-section 2.4).

**Example 2.4.1.** Consider the initial value problem

$$\begin{cases} y'(t) = y(t) 0 \leq t \leq T \\ y(0) = \left(\underline{y}(0, \quad \alpha), \; \bar{y}(0, \quad \alpha)\right). \end{cases}$$

by using the fuzzy Laplace transform method, we have

$L[y'(t)] = L\;[y(t)]$, and $L[y'(t)] = \int_0^\infty y'(t) \odot e^{-pt}dt$ in (i)-differentiable then by using Case (i), we have $L[y'(t)] = (s\,L\;[y(t)]) \ominus y(0)$

Therefore, $L[y(t)] = s\,L\;[y(t)]) \ominus y(0)$

$$\left.\begin{aligned} l[\bar{y}(t,\ \alpha)] &= s\,l\left[\underline{y}(t,\ \alpha)\right] - \underline{y}(0,\alpha) \\ l\left[\underline{y}(t,\ \alpha)\right] &= s\,l[\bar{y}(t,\ \alpha)] - \bar{y}(0,\alpha) \ldots \end{aligned}\right\} \tag{17}$$

Hence, the solution of system (18) is:

$$l[\bar{y}(t,\ \alpha)] = -\bar{y}(0,\alpha)\left(\frac{s}{s^2-1}\right) + \underline{y}(0,\alpha)\left(-\frac{1}{s^2-1}\right)$$

$$l\left[\underline{y}(t,\ \alpha)\right] = -\underline{y}(0,\alpha)\left(\frac{s}{s^2-1}\right) + \bar{y}(0,\alpha)\left(-\frac{1}{s^2-1}\right)$$

Thus

$$\bar{y}(t,\alpha) = -\bar{y}(0,\alpha)l^{-1}\left[\left(\frac{s}{s^2-1}\right)\right] + \underline{y}(0,\alpha)l^{-1}\left[\left(-\frac{1}{s^2-1}\right)\right]$$

$$\underline{y}(t,\alpha) = -\underline{y}(0,\alpha)l^{-1}\left[\left(\frac{s}{s^2-1}\right)\right] + \bar{y}(0,\alpha)l^{-1}\left[\left(-\frac{1}{s^2-1}\right)\right]$$

Finally, we have:

$$\bar{y}(t,\alpha) = e^{-t}\left(\frac{\underline{y}(0,\ \alpha) - \bar{y}(0,\ \alpha)}{2}\right) - e^{t}\left(\frac{\underline{y}(0,\ \alpha) + \bar{y}(0,\ \alpha)}{2}\right)$$

$$\underline{y}(t,r) = e^{-t}\left(\frac{-\underline{y}(0,\ \alpha) + \bar{y}(0,\ \alpha)}{2}\right) - e^{t}\left(\frac{\underline{y}(0,\ \alpha) + \bar{y}(0,\ \alpha)}{2}\right)$$

If $y'(t)$ in (ii)-is differentiable, then by using Case II, we have $L[y'(t)] = -(y(0)) \ominus (-s\,L\,[y(t)])$. Therefore, $L[y(t)] = (-y(0)) \ominus (-s\,L\,[y(t)])$

$$\left.\begin{aligned} l[\bar{y}(t,\ \alpha)] &= s\,l[\bar{y}(t,\ \alpha)] - \bar{y}(0,\alpha) \\ l\left[\underline{y}(t,\ \alpha)\right] &= sl\left[\underline{y}(t,\ \alpha)\right] - \underline{y}(0,\alpha) \ldots \end{aligned}\right\} \tag{18}$$

Hence, the solution of system (2) is:

$$l[\bar{y}(t,\ \alpha)] = -\bar{y}(0,\alpha)\left(\frac{1}{1+s}\right)$$

$$l\left[\underline{y}(t,\ \alpha)\right] = -\underline{y}(0,\alpha)\left(\frac{1}{1+s}\right)$$

Thus

$$\bar{y}(t,\alpha) = -\bar{y}(0,\alpha)l^{-1}\left(\frac{1}{1+s}\right)$$

$$\underline{y}(t,\alpha) = -\underline{y}(0,\alpha)l^{-1}\left(\frac{1}{1+s}\right)$$

Finally, we have:

$$\bar{y}(t,\alpha) = -\bar{y}(0,\alpha)e^{-t}$$
$$\underline{y}(t,\alpha) = -\underline{y}(0,\alpha)e^{-t}$$

## 4. Conclusion

This chapter discussed the Laplace transform and fuzzy Laplace transform in dynamic systems. Using the Laplace transform and the fuzzy Laplace transform methods, we provided solutions to first- and second-order ordinary differential equations and first- and second-order fuzzy ordinary differential equations. We demonstrated how the Laplace transform method could determine the solution (current flow) of first- and second-order electrical circuits and a mechanical system's solution (vibration frequency).

## Author details

Shivram Sharma[1], Praveen Kumar Sharma[2*] and Jitendra Kaushik[3]

1 Department of Mathematics, Govt. PG College, Guna, MP, India

2 Department of Mathematics, SVIS, Shri Vaishnav Vidyapeeth Vishwavidyalaya, Indore, MP, India

3 MIT Art, Design and Technology University, Pune, India

*Address all correspondence to: praveen_jan1980@rediffmail.com

## References

[1] Dass HK, Verma R. Higher Engineering Mathematics. New Delhi: S. Chand Publication

[2] Duffy DG. Transform Methods for Solving Partial Differential Equations. USA: CRC Press; 1994

[3] Franklin G, Powell D, Emami-Naeini A. Feedback Control of Dynamic Systems. USA: Prentice-Hall; 2002

[4] Rosenbrock HH. State-Space and Multivariable Theory. UK: Nelson; 1970

[5] Ramanna BV. Higher Engineering Mathematics. India: Tata McGraw Hill Publication

[6] Zadeh LA. Fuzzy sets. Information Controlled. 1965;**8**:338-353

[7] Najariyan M, Farahi MH. A new approach for the optimal fuzzy linear time-invariant controlled system with fuzzy coefficients. Journal of Computational and Applied Mathematics. 2014;**259**:682-694

[8] Najariyan M, Farahi MH. A new approach for solving a class of fuzzy optimal control systems under generalized Hukuhara differentiability. Journal of the Franklin Institute. 2015; **352**(5):1836-1849

[9] Najariyan M, Zhao Y. Fuzzy fractional quadratic regulator problem under granular fuzzy fractional derivatives. IEEE Transactions on Fuzzy Systems. 2017;**26**(4):2273-2288

[10] Mazandarani M, Pariz N. Sub-optimal control of fuzzy linear dynamical systems under granular differentiability concept. ISA Transactions. 2018;**76**: 1-17

[11] Najariyan M, Zhao Y. On the stability of fuzzy linear dynamical systems. Journal of the Franklin Institute. 2020;**357**(9):5502-5522

[12] Allahviranloo T, Abbasbandy S, Salahshour S, Hakimzadeh A. A new method for solving fuzzy linear differential equations. Computing. 2011; **59**(3):181-197

[13] Salahshour S, Allahviranloo T. Applications of fuzzy Laplace transforms. Soft Computing. 2013;**17**(1): 145-158

[14] Chang SSL, Zadeh L. On fuzzy mapping and control. IEEE Transactions on Systems and Cybernetics. 1972;**2**: 30-34

[15] Kandel A. Fuzzy dynamical systems and the nature of their solutions. In: Wang PP, Chang SK, editors. Fuzzy Sets Theory and Application to Policy Analysis and Information Systems. New York: Plenum Press; 1980. pp. 93-122

[16] Kandel A, Byatt WJ. Fuzzy differential equations. In: Proceedings of the International Conference on Cybernetics and Society. Tokyo; 1978. pp. 1213-12160

[17] Sharma PK, Sharma S, Kaushik J, Goyal P. A Book chapter with the title "Applications of fuzzy set and fixed point theory in Dynamical systems" under an Open Access Book with the title. In: Qualitative and Computational Aspects of Dynamical Systems. UK: Intech Open; 2022

Chapter 2

# Numerical Methods: Euler and Runge-Kutta

*Victor Akinsola*

## Abstract

Most real life phenomena change with time, hence dynamic. Differential equations are used in mathematical modeling of such scenarios. Linear differential equations can be solved analytically but most real life applications are nonlinear. Numerical solutions of nonlinear differential equations are approximate solutions. Euler and Runge-Kutta method of order four are derived, explained and illustrated as useful numerical methods for solving single and systems of linear and nonlinear differential equations. Accuracy of a numerical method depends on the step size used and degree of nonlinearity of the equations. Stiffness is another challenge with numerical solutions of nonlinear differential equations. Although better accuracy can be obtained with smaller step size, this takes more computational effort and time. Algorithms and codes can be written using available computer programming software to overcome this challenge and to avoid computational error. The Runge-Kutta method is more applicable and accurate for diverse classes of differential equations.

**Keywords:** numerical solution, Euler method, Runge-Kutta method of order four (RK4), accuracy

## 1. Introduction

In numerical analysis, mathematical methods are employed to produce numerical answers to mathematical expressions. It entails developing, examining, and putting into practice computer algorithms to solve continuous mathematics problems numerically [1]. These problems may originate from real-life applications in the natural sciences, social sciences, engineering, medical sciences and business where variables which vary continuously are involved.

Differential equations are the foundation of the majority of mathematical models used in the natural sciences and engineering. The numerical technique finds the solution function at a discrete set of points, approximating the derivatives or integrals in the relevant equation.

In an effort to obtain solutions that are more precise, more affordable, or more resilient to instabilities in the issue data, a number of techniques have been developed. Many of the differential equations that arise in practical problems are usually nonlinear. Nonlinear differential equations are nearly impossible to be solved precisely [2]. Most differential equations that arise in practical problems are typically nonlinear. The computational labour of solving a system of first order equations may be formidable even when the exact solution is possible [3]. It is for these reasons that techniques have been developed for computing to any degree of accuracy the numerical solution of almost any of such problems.

These techniques typically fall into "families" of increasing order of accuracy, with Euler's method (or a close relative) frequently making up the lowest order. Adams-Bashforth techniques are "multistep" methods, whereas Runge-Kutta methods are "single-step" methods. These methods have been very successful and reliable.

In this chapter, the derivation of Euler and Runge Kutta methods are explained and illustrative examples given to explain how they are used. Algorithms/codes of the examples written with Maple mathematical programming software were also included for better comprehension and further practice.

## 2. The Euler method

The popular Euler method published in 1768 is attributed to Leonhard Euler (1707–1783). The method is one of the most straightforward and fundamental method of solving single and systems of ordinary differential eqs. [1]. The basic idea is as follows;

Consider initial value system:

$$\frac{dy}{dt} = f(t, x, y) y(0) = y_0 \tag{1}$$

$$\frac{dx}{dt} = g(t, x, y) x(0) = x_0 \tag{2}$$

By the definition of a derivative,

$$y'(t) = \lim_{h \to 0} \frac{f(t+h) - f(t)}{h} \tag{3}$$

$$x'(t) = \lim_{h \to 0} \frac{g(t+h) - f(t)}{h} \tag{4}$$

for small $h > 0$, then, Eqs. (3) and (4) imply that a reasonable difference approximation for $y'(t)$ and $x'(t)$ [4] are.

$$y'(t) = \frac{f(t+h) - f(t)}{h} \tag{5}$$

$$x'(t) = \frac{g(t+h) - f(t)}{h} \tag{6}$$

Substituting Eq. (5) and Eq. (6) into Eq. (1) and Eq. (2) respectively yield the difference equations

$$\frac{f(t+h) - f(t)}{h} = f(t, x, y) \tag{7}$$

$$\frac{g(t+h) - f(t)}{h} = g(t, x, y) \tag{8}$$

which approximates the differential Eqs. (1) and (2).

According to [1], the Euler's method approximates a solution (x, y) with step size $h$ by:

$$y_{k+1} = y_k + hf(x_k, y_k) \tag{9}$$

$$x_{k+1} = x_k + hg(x_k, y_k) \tag{10}$$

Expressing Eq. (1) and Eq. (2) in vector notation:

$$\frac{\mathrm{dY}}{\mathrm{dt}} = F(Y), Y_0 \tag{11}$$

Where $Y = (y, x)$, $Y_0 = (y_0, x_0)$, $F(Y) = (f(x,\ y), g(x, y))$ and $\frac{\mathrm{dY}}{\mathrm{dt}} = \left(\frac{dy}{dt}, \frac{dx}{dt}\right)$.

The Euler's method approximates a solution (x, y) by:

$$(x_{k+1}, y_{k+1}) = (x_k, y_k) + hF(x_k, y_k) \tag{12}$$

**Illustration 1**

Given the initial value problem in [5]:

$$y' = \frac{y - x}{y + x}, y(0) = 1.$$

Determine.

$y$ for $x = 0.1$, using the Euler method with the step size $h = 0.02$.

**Solution**

The Euler formula is.

$y_{n+1} = y_n + hf(x_n, y_n)$ where $n$ is the number of iteration.

For the first iteration which is $n = 0$, the Euler formula becomes

$$y_1 = y_0 + hf(x_0, y_0)$$

$x_0 = 0$ and $y_0 = 1$ as the given initial condition.

$$x_1 = x_0 + h$$

$$x_n = x_0 + nh$$

$$n = \frac{x_n - x_0}{h} = \frac{0.1 - 0}{0.02} = 5$$

$$y_1 = 1 + 0.02f(0, 1) = 1 + 0.02\left(\frac{1 - 0}{1 + 0}\right) = 1 + 0.02 = 1.02$$

When $n = 1$, $x_1 = x_0 + h = 0 + 0.02 = 0.02$.

$y_2 = y_1 + 0.02f(x_1, y_1) = 1.02 + 0.02f(0.02, 1.02) = 1.02 + 0.02\left(\frac{1.02 - 0.02}{1.02 + 0.02}\right) =$ 1.039230769 When $n = 2$, $x_2 = x_0 + 2h = 0 + 2(0.02) = 0.04$

$$y_3 = y_2 + 0.02f(x_2, y_2) = 1.039230769 + 0.02\left(\frac{1.039230769 - 0.04}{1.039230769 + 0.04}\right) = 1.057748232$$

When $n = 3$, $x_3 = x_2 + h = 0.04 + 0.02 = 0.06$

| X | y(x) at h = 0.02, n = 5 | y(x) at h = 0.01, n = 10 | y(x) at h = 0.005, n = 20 |
|---|---|---|---|
| 0 | 1.0 | 1.0 | 1.0 |
| 0.02 | 1.02000000 | 1.019803922 | 1.019706799 |
| 0.04 | 1.039230769 | 1.038852886 | 1.038665609 |
| 0.06 | 1.057748232 | 1.057200704 | 1.056929219 |
| 0.08 | 1.075601058 | 1.074894334 | 1.074543762 |
| 0.10 | 1.092831936 | 1.091975065 | 1.091549847 |

**Table 1.**
*Numerical solution of Example on Euler's method.*

$$y_4 = y_3 + 0.02f(x_3, y_3) = 1.057748232 + 0.02\left(\frac{1.057748232 - 0.06}{1.057748232 + 0.06}\right) = 1.075601058$$

When $n = 4, x_4 = x_3 + h = 0.06 + 0.02 = 0.08$

$$y_5 = y_4 + 0.02f(x_4, y_4) = 1.075601058 + 0.02\left(\frac{1.075601058 - 0.08}{1.075601058 + 0.08}\right) = 1.092831936$$

**Table 1** gives the numerical solution of Illustration 1 using different step sizes. Obviously this takes more computational effort since more iteration is needed but gives more accurate result.

The algorithm/ code given below can be adapted to various single differential equations for the Euler method (**Figure 1**).

**Algorithm/ Codes using Maple programming software**

```
## Illustration on Euler method#
ode:=diff(y(x),x)=(y(x)-x)/(y(x)+x);
sol:dsolve({ode,y(0)=1},y(x));
##restart:y:=array(0..200):x:=array(0..200):n:=5.0:A:=0.0:h:=0.02:x[0]:=
0.0:y[0]:=1.0:
for m from 0 to n do x[m]:=A+m*h:y[0]:=1.0:od:
for m from 0 to n do y[m+1]:=y[m]+h*((y[m]-x[m])/(y[m]+x[m])):od:
for m from 0 to n do print (m, x[m], y[m]);od:
0, 0., 1.0
1, 0.02, 1.020000000
2, 0.04, 1.039230769
3, 0.06, 1.057748232
4, 0.08, 1.075601058
5, 0.10, 1.092831936
```

For more on Maple programming codes and algorithms see [6].

**Illustration 2**

Consider the Lorenz dynamical system given as.

$$\frac{dx}{dt} = \sigma(y(t) - x(t)), x(0) = 0.1$$

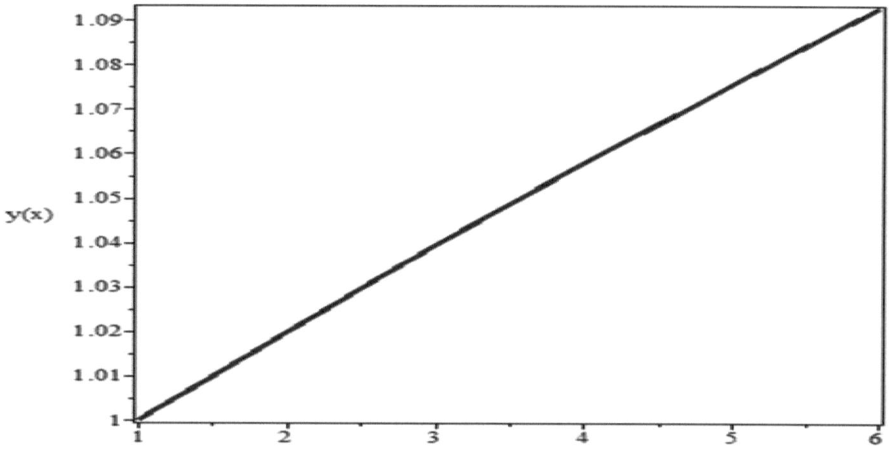


**Figure 1.**
*Graph of numerical solution of Illustration 1 for six iterations.*

$$\frac{dy}{dt} = -x(t)z(t) + rx(t) - y(t), y(0) = 0.1$$

$$\frac{dz}{dt} = x(t)y(t) - bz(t), z(0) = 0.1$$

Where $\sigma = 10$, $r = 28$ and $b = 8/3$ with step size 0.01 determine the solution at $t = 10$ using the Euler's method.

**Solution**

The Lorenz system is a classical nonlinear dynamical system. Due to the high degree of nonlinearity of the system and the number of iterations required, its numerical solution is only easily tractable using computer programming. The algorithm/Code in Maple mathematical software is given below after the table of numerical solution (**Figures 2–4**) and (**Table 2**).

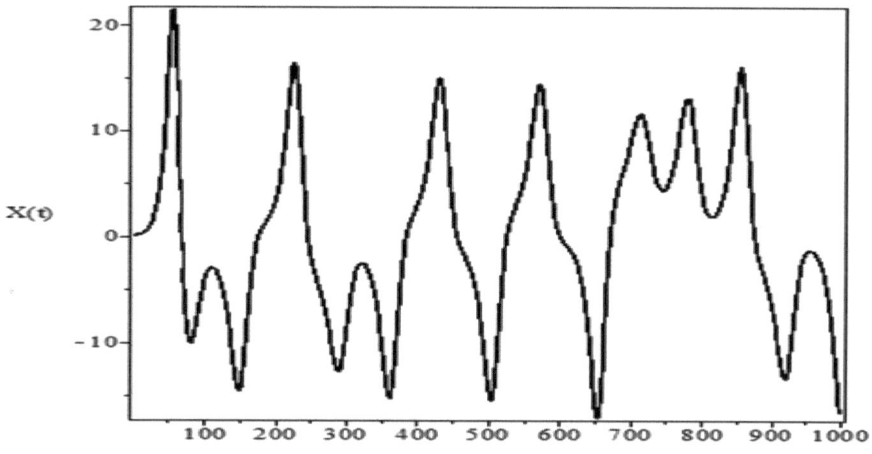


**Figure 2.**
*Graphical representation of numerical solution of Lorenz system $X(t)$ using Euler method for 1000 iterations as in **Table 2**.*

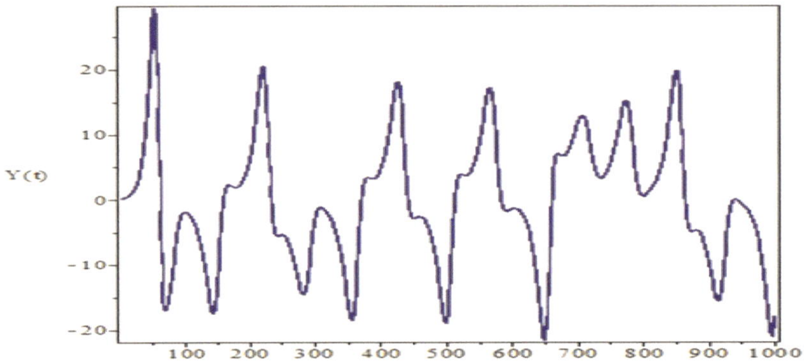


**Figure 3.**
*Graphical representation of numerical solution of Lorenz system Y(t) using Euler method for 1000 iterations as in* ***Table*** *2.*

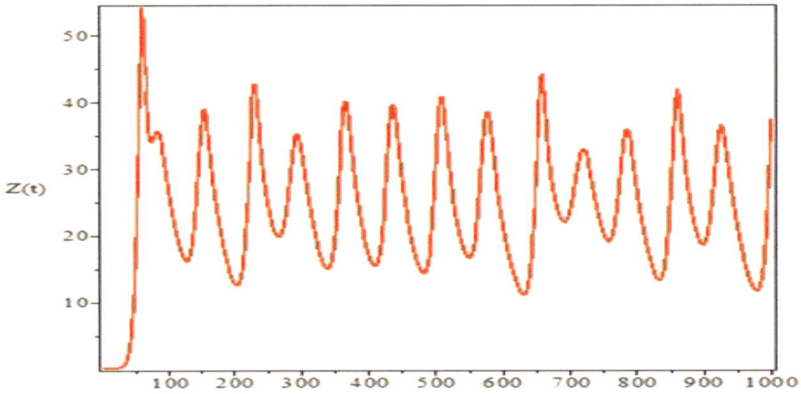


**Figure 4.**
*Graphical representation of numerical solution of Lorenz system Z(t) using Euler method for 1000 iterations as in* ***Table*** *2.*

| *t* | X(t) | Y(t) | Z(t) |
|---|---|---|---|
| 1.0 | −4.277716783 | −1.835441598 | 26.77337837 |
| 2.0 | 3.307860445 | 5.707928748 | 12.82481891 |
| 3.0 | −9.048257350 | −3.394242770 | 33.60229026 |
| 4.0 | 2.906631532 | 4.606304372 | 17.27981641 |
| 5.0 | −14.25707158 | −18.85653324 | 29.67196074 |
| 6.0 | 0.1457099124 | −1.419066178 | 21.71150940 |
| 7.0 | 8.807447039 | 11.68841199 | 23.40171689 |
| 8.0 | 4.038757258 | 0.5104950902 | 27.19520327 |
| 9.0 | −6.260708216 | −9.461245716 | 18.53277577 |
| 10.0 | −16.65763702 | −18.00805550 | 37.42332818 |

**Table 2.**
*Numerical solution of the Lorenz system using Euler method at step size 0.01.*

$$t_n = t_0 + nh$$

$$n = \frac{t_n - t_0}{h} = \frac{10 - 0}{0.01} = 1000$$

---

**Algorithm/Maple programming Code of the numerical solution of Lorenz system using Euler's method**

```
## This is the numerical solution of Lorenz system using Euler method##
sys:=diff(x(t),t)=sigma*(y(t)-x(t)),diff(y(t),t)=-x(t)*z(t)+r*x(t)-y(t),
diff(z(t),t)=x(t)*y(t)-b*z(t);
##restart:X:=array(0..2000000):Y:=array(0..2000000):Z:=array
(0..2000000):T:=array(0..2000000):A:=0.0:h:=0.01:N:=1000:sigma:=
10.0:b:=(8/3):r:=28.0: for m from 0 to N do T[m]:=A+m*h:X[0]:=0.1:
Y[0]:=0.1:Z[0]:=0.1:od: for m from 0 to N do X[m+1]:=X[m]+
h*(sigma*(Y[m]-X[m])):Y[m+1]:=Y[m]+h*(-X[m]*Z[m]+r*X[m]-Y[m]):
Z[m+1]:=Z[m]+h*(X[m]*Y[m]-b*Z[m]):od:
for m from 0 by 100 to N do print (m, T[m], X[m],Y[m],Z[m]);od: with(plots):
func1:=listplot([seq(X[m],m=0..N)],style=line,color=black,thickness=1):
display(func1,labels=["t","X(t)"],axes=boxed,caption="Figure1:Euler solution for
X(t)");
func2:=listplot([seq(Y[m],m=0..N)],style=line,color=blue,thickness=1):
display(func2,labels=["t","Y(t)"],axes=boxed,caption="Figure2:Euler solution for
Y(t)"); func3:=listplot([seq(Z[m],m=0..N)],style=line,color=red, thickness=1):
display(func3,labels=["t","Z(t)"],axes=boxed,caption="Figure
3:Euler solution for Z(t)");
```

---

For more on Lorenz system see [7].

## 3. The Runge-Kutta method of order four

One of the well-known numerical techniques for solving differential equations is the Runge-Kutta method. The Runge-Kutta method is a family of methods which depend on the order of derivatives involved. The Runge-Kutta method of order four is the classical form of the method in which four values of the derivatives are used for each iteration [1].

### 3.1 Derivation of Runge-Kutta method of order four

The Taylor series of functions of one variable is:

$$y(x) = y(x_0) + y'(x_0)(x - x_0) + \frac{y''(x_0)}{2!}(x - x_0)^2 + \frac{y'''(x_0)}{3!}(x - x_0)^3 + \cdots \quad (13)$$

$$y(x) = \sum_{n=0}^{\infty} \frac{y^{(n)}}{n!}(x - x_0)^n \quad (14)$$

when $x = x_0 + h$, $x - x_0 = h$.

Then Eq. (13) becomes.

$$y(x) = y(x_0) + hy'(x_0) + h^2\frac{y''(x_0)}{2!} + h^3\frac{y'''(x_0)}{3!} + \cdots \tag{15}$$

Taylor series of functions of two variables is:

$$f(x,y) = \sum_{i=0}^{\infty}\frac{1}{n!}\left\{(x-x_0)\frac{\partial}{\partial x} + (y-y_0)\frac{\partial}{\partial y}\right\}^i f(x,y) \tag{16}$$

when $x = x_0 + mh$, $x - x_0 = mh$and $y - y_0 = nh$

$$f(x,y) = \sum_{i=0}^{\infty}\frac{1}{n!}\left\{mh\frac{\partial}{\partial x} + nh\frac{\partial}{\partial y}\right\}^i f(x,y) \tag{17}$$

$$\begin{aligned} f(x.y) =& f(x_0,y_0) + f'(x_0,y_0)(mh+nh) + \frac{f''(x_0,y_0)(mh+nh)^2}{2!} \\ &+ \frac{f'''(x_0,y_0)(mh+nh)^3}{3!} + \cdots \end{aligned} \tag{18}$$

$$\begin{aligned} f(x.y) =& f(x_0,y_0) + mhf'(x_0,y_0) + nhf'(x_0,y_0) + \frac{f''(x_0,y_0)\left((mh)^2 + 2(mh)(nh) + (nh)^2\right)}{2!} \\ &+ \frac{f'''(x_0,y_0)\left((mh)^3 + 3(mh)^2(nh) + 3(mh)(nh)^2 + (nh)^3\right)}{3!} + \cdots \end{aligned} \tag{19}$$

$$\begin{aligned} f(x.y) =& f(x_0,y_0) + h\left[mf_x + nf_y\right] + \frac{h^2}{2!}\left[m^2f_{xx} + 2mnf_{xy} + n_2f_{yy}\right] \\ &+ \frac{h^3}{3!}\left[m^3f_{xxx} + 3m^2nf_{xxy} + 3mn^2f_{xyy} + n^3f_{yyy}\right] + \cdots \end{aligned} \tag{20}$$

where the partial derivatives are all evaluated at the point $(x_0, y_0)$. Consider the differential equation

$$\frac{dy}{dx} = y' = f(x,y) \tag{21}$$

$$df = \frac{\partial f}{\partial x}dx + \frac{\partial f}{\partial y}dy \tag{22}$$

$$\frac{d^2f}{dx^2} = \frac{df}{dx} = f' = \frac{\partial f}{dx}\frac{dx}{dx} + \frac{\partial f}{\partial y}\frac{dy}{dx} \tag{23}$$

$$y'' = f' = \frac{\partial f}{dx} + \frac{\partial f}{\partial y}\frac{dy}{dx} \tag{24}$$

$$y'' = f' = \frac{\partial f}{dx} + \frac{\partial f}{\partial y}f \tag{25}$$

$$y'' = f' = f_x + ff_y \tag{26}$$

$$y''' = f'' = \left[f_x + ff_y\right]' = \left[f_x\right]' + \left[ff_y\right]' \tag{27}$$

*DOI: http://dx.doi.org/10.5772/intechopen.108533*

$$(f_x)' = f_x + ff_{xy} \tag{28}$$

$$\left(ff_y\right)' = f\left(f_y'\right) + f_y f' \tag{29}$$

$$y''' = f'' = f_{xx} + 2ff_{xy} + f^2 f_{yy} + f_x f_y + ff_y{}^2 \tag{30}$$

From Eq. (15).

$$y(x) - y(x_0) = hy'(x_0) + h^2\frac{y''(x_0)}{2!} + h^3\frac{y'''(x_0)}{3!} + \cdots \tag{31}$$

Hence

$$y(x) - y(x_0) = hf(x_0) + \frac{h^2}{2!}\left(f_x + ff_y\right) + \frac{h^3}{3!}\left(f_{xx} + 2ff_{xy} + f^2 f_{yy} + f_x f_y + ff_y{}^2\right) + \cdots \tag{32}$$

The main idea is to select several points so that the Taylor series of expansion Eq. (19), coincide with the terms on the right hand side of Eq. (32).

$$\text{Suppose } m_1 = hf(x_0, y_0) \tag{33}$$

$$m_2 = hf(x_0 + nh, y_0 + nm_1) \tag{34}$$

$$m_3 = hf(x_0 + ph, y_0 + pm_2) \tag{35}$$

$$m_4 = hf(x_0 + qh, y_0 + qm_3) \tag{36}$$

Using Eq. (32), these values become;

$$m_1 = hf \tag{37}$$

$$m_2 = h\left[f + nh\left(f_x + ff_y\right) + \frac{(nh)^2}{2}\left(f_{xx} + 2ff_{xy} + f^2 f_{yy}\right) + \cdots\right] \tag{38}$$

$$m_3 = h\left[f + ph\left(f_x + ff_y\right) + \frac{(ph)^2}{2}\left(f_{xx} + 2ff_{xy} + f^2 f_{yy}\right) + 2np\left(f_x f_y + ff_y{}^2\right) + \cdots\right] \tag{39}$$

$$m_4 = h\left[f + qhc + \frac{(qh)^2}{2}\left(f_{xx} + 2ff_{xy} + f^2 f_{yy}\right) + 2pq\left(f_x f_y + ff_y{}^2\right) + \cdots\right] \tag{40}$$

where all functions are evaluated at the point ($x_0$, $y_0$).
Considering an expression of the form

$$am_1 + bm_2 + cm_3 + dm_4 \tag{41}$$

Equating the expression to the right hand side of Eq. (37) through Eq. (40) to obtain four equations in seven unknowns.

$$\textit{Coefficient of } hf : a + b + c + d = 1 \tag{42}$$

$$\textit{Coefficient of } hf\left(f_x f_y + f f_y^{\,2}\right)\ bn + cp + dq = \frac{1}{2} \tag{43}$$

$$\text{Coefficient of } hf\left(f_{xx} + 2ff_{xy} + f^2 f_{yy}\right) : bn^2 + cp^2 + dq^2 = \frac{1}{3} \tag{44}$$

$$\textit{Coefficient of } hf\left(f_x f_y + f f_y^{\,2}\right) : cnp + dpq = \frac{1}{6} \tag{45}$$

Choosing three unknown quantities arbitrarily since there are four equations in seven unknowns.

Let $n = p = \frac{1}{2}$ and $q = 1$ in Eq. (42) through Eq. (45). Then

$$a + b + c + d = 1 \tag{46}$$

$$b + c + 2d = 1 \tag{47}$$

$$3b + 3c + 12d = 4 \tag{48}$$

$$3c + 6d = 2 \tag{49}$$

Solving Eq. (46) through Eq. (47) simultaneously produced.

$$a = d = \frac{1}{6}, b = c = \frac{1}{3}$$

Connecting Eq. (33) and the expression (41).

$$y(x) - y(x_0) = am_1 + bm_2 + cm_3 + dm_4 \tag{50}$$

$$y(x) - y(x_0) = \frac{m_1}{6} + \frac{m_2}{3} + \frac{m_3}{3} + \frac{m_4}{6} \tag{51}$$

$$y(x) = y(x_0) + \frac{1}{6}(m_1 + 2m_2 + 2m_3 + m_4) \tag{52}$$

In general

$$y_{n+1} = y_n + \frac{1}{6}(m_1 + 2m_2 + 2m_3 + m_4) \tag{53}$$

Where

$$m_1 = hf(x_n, y_n)$$

$$m_2 = hf\left(x_n + \frac{1}{2}h, y_n + \frac{1}{2}m_1\right)$$

$$m_3 = hf\left(x_n + \frac{1}{2}h, y_n + \frac{1}{2}m_2\right)$$

$$m_4 = hf(x_n + h, y_n + m_3)$$

This is the Runge-Kutta formula of order four for a single first order differential equation.

The formula can be generalized for system of differential equations.

Consider the system of two initial-value differential equation:

$$\frac{dx}{dt} = f(t, x, y)x(0) = y_0 \tag{54}$$

$$\frac{dy}{dt} = g(t, x, y)y(0) = x_0 \tag{55}$$

The Runge-Kutta formula of order four for system of two initial-value differential equation of the form of Eq. (54) and Eq. (55) is

$$x_{n+1} = x_n + \frac{1}{6}(k_1 + 2k_2 + 2k_3 + k_4) \tag{56}$$

$$y_{n+1} = y_n + \frac{1}{6}(m_1 + 2m_2 + 2m_3 + m_4) \tag{57}$$

Where

$$k_1 = hf(t_n, x_n, y_n)$$
$$m_1 = hg(t_n, x_n, y_n)$$
$$k_2 = hf\left(t_n + \frac{1}{2}h, x_n + \frac{1}{2}k_1, y_n + \frac{1}{2}m_1\right)$$
$$m_2 = hg\left(t_n + \frac{1}{2}h, x_n + \frac{1}{2}k_1, y_n + \frac{1}{2}m_1\right)$$
$$k_3 = hf\left(t_n + \frac{1}{2}h, x_n + \frac{1}{2}k_2, y_n + \frac{1}{2}m_2\right)$$
$$m_3 = hg\left(t_n + \frac{1}{2}h, x_n + \frac{1}{2}k_2, y_n + \frac{1}{2}m_2\right)$$
$$k_4 = hf(t_n + h, x_n + k_3, y_n + m_3)$$
$$m_4 = hg(t_n + h, x_n + k_3, y_n + m_3)$$

Generalizations of the formular to system of more than two equations follow a similar pattern.

Further insights and resources on the Runge- Kutta method can be found in [8–10].

**Illustration 3**

Given the initial value problem.

$$y' = x + y, y(0) = 1..$$

Determine.

$y$ for $x = 0.2$, using the Runge-Kutta method of order four (RK4) with the step size $h = 0.1$.

**Solution**

The RK4 formula is:

$$y_{n+1} = y_n + \frac{1}{6}(m_1 + 2m_2 + 2m_3 + m_4)$$

Where

$$m_1 = hf(x_n, y_n)$$

$$m_2 = hf\left(x_n + \frac{1}{2}h, y_n + \frac{1}{2}m_1\right)$$

$$m_3 = hf\left(x_n + \frac{1}{2}h, y_n + \frac{1}{2}m_2\right)$$

$$m_4 = hf(x_n + h, y_n + m_3)$$

$$x_n = x_0 + nh$$

$$n = \frac{x_n - x_0}{h} = \frac{0.2 - 0}{0.1} = 2$$

Hence two iterations shall be needed.
When $n = 0$

$$y_1 = y_0 + \frac{1}{6}(m_1 + 2m_2 + 2m_3 + m_4)$$

$$m_1 = hf(x_0, y_0) = 0.1f(0, 1) = 0.1(0 + 1) = 0.1$$

$$m_2 = hf\left(x_0 + \frac{1}{2}h, y_0 + \frac{1}{2}m_1\right) = 0.1f\left(0 + \frac{1}{2}(0.1), 1 + \frac{1}{2}(0.1)\right)$$
$$= 0.1f(0.05, 1.05) = 0.1(0.05 + 1.05) = 0.1(1.1) = 0.11$$

$$m_3 = hf\left(x_0 + \frac{1}{2}h, y_0 + \frac{1}{2}m_2\right) = 0.1f\left(0 + \frac{1}{2}(0.1), 1 + \frac{1}{2}(0.11)\right)$$
$$= 0.1f(0.05, 1.055) = 0.1(0.05 + 1.055) = 0.1(1.105) = 0.1105$$

$$m_4 = hf(x_0 + h, y_0 + m_3) = 0.1f(0 + 0.1, 1 + 0.1105)$$
$$= 0.1f(0.1, 1.1105) = 0.1(0.1 + 1.1105) = 0.1(1.2105) = 0.12105$$

$$y_1 = y_0 + \frac{1}{6}(m_1 + 2m_2 + 2m_3 + m_4)$$
$$= 1 + \frac{1}{6}(0.1 + 2(0.11) + 2(0.1105) + 0.12105) = 1.110342$$

For the second iteration, when n = 1:

$$y_2 = y_1 + \frac{1}{6}(m_1 + 2m_2 + 2m_3 + m_4)$$

Where

$$x_1 = x_0 + h = 0 + 0.1 = 0.1$$

$$m_1 = hf(x_1, y_1) = 0.1f(0.1, 1.110342) = 0.1(0.1 + 1.110342)$$
$$= 0.1(1.210342) = 0.1210342$$

$$m_2 = hf\left(x_1 + \frac{1}{2}h, y_1 + \frac{1}{2}m_1\right) = 0.1f\left(0.1 + \frac{1}{2}(0.1), 1.110342 + \frac{0.1210342}{2}\right)$$
$$= 0.1f(0.15, 1.1.708591) = 0.13208591$$

$$m_3 = 0.1f(0.15, 1.176385) = 0.1326385$$

$$m_4 = 0.1f(0.2, 1.2429805) = 0.144298049$$

$$y_2 = 1.110342 + \frac{1}{6}(0.1210342 + 2(0.13208591) + 2(0.1326385) + 0.144298049)$$
$$= 1.242805512$$

**Illustration 4**

Consider the Lorenz dynamical system given as.

$$\frac{dx}{dt} = \sigma(y(t) - x(t)), x(0) = 0.1.$$
$$\frac{dy}{dt} = -x(t)z(t) + rx(t) - y(t), y(0) = 0.1.$$
$$\frac{dz}{dt} = x(t)y(t) - bz(t), z(0) = 0.1.$$

Where $\sigma = 10$, $r = 28$ and $b = 8/3$ with step size 0.1 determine the numerical solution at $t = 10$ using the method of RK4 (**Figures 5–7**).

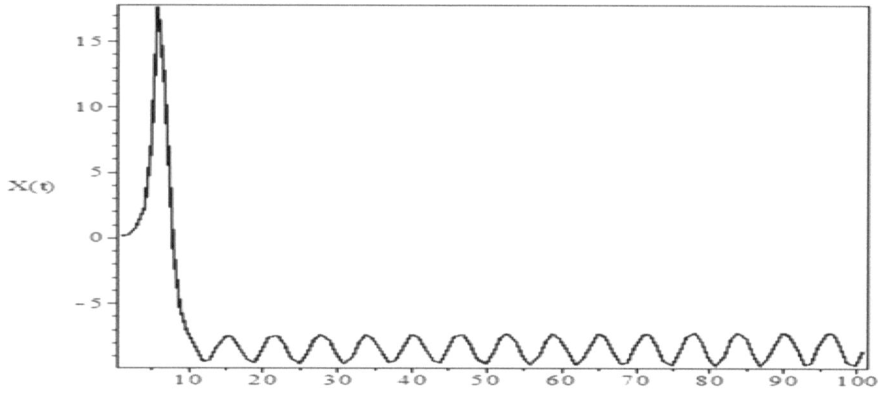


**Figure 5.**
*Graphical representation of numerical solution of Lorenz system X(t) using Runge-Kutta method of order four for 100 iterations as in* ***Table 3****.*

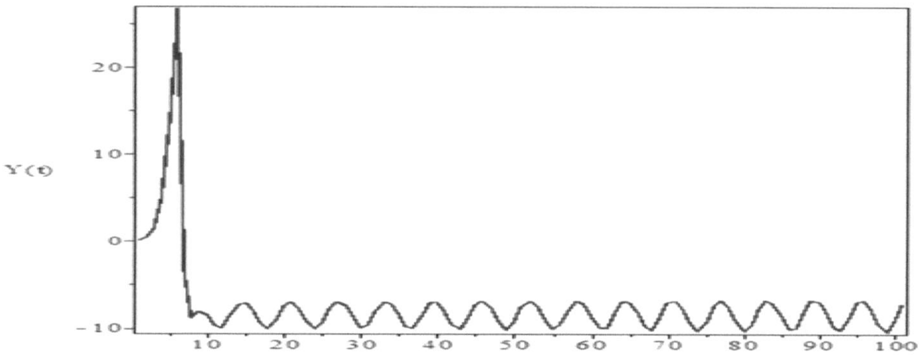


**Figure 6.**
*Graphical representation of numerical solution of Lorenz system Y(t) using Runge-Kutta method of order four for 100 iterations as in* ***Table 3****.*

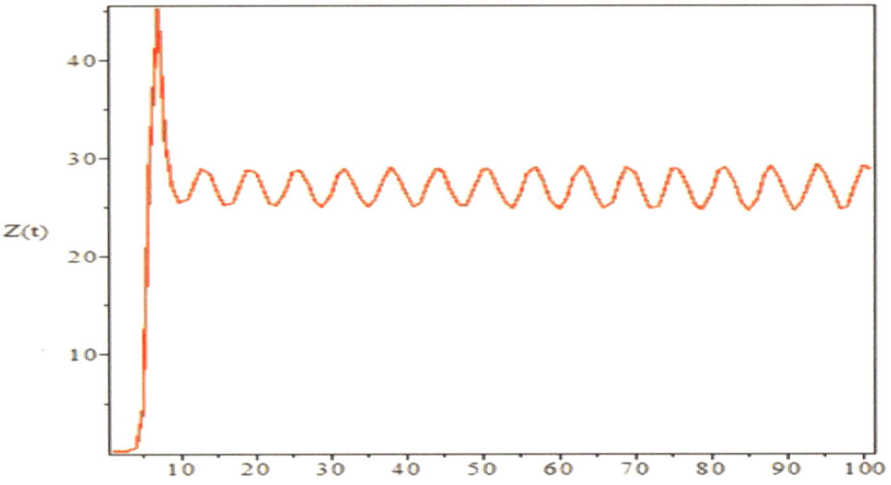


**Figure 7.**
*Graphical representation of numerical solution of Lorenz system X(t) using Runge-Kutta method of order four for 100 iterations as in **Table 3**.*

| *t* | **X(t)** | **Y(t)** | **Z(t)** |
|---|---|---|---|
| 1.0 | −8.437853127 | −9.711139533 | 25.75673332 |
| 2.0 | −7.628715786 | −6.966090160 | 26.86095558 |
| 3.0 | −9.574336573 | −9.690236844 | 28.21573058 |
| 4.0 | −7.524494281 | −8.107038156 | 24.99141661 |
| 5.0 | −8.752300839 | −7.562662622 | 28.79099030 |
| 6.0 | −8.804061170 | −9.973397587 | 25.79606864 |
| 7.0 | −7.396786248 | −6.824809818 | 26.47905713 |
| 8.0 | −9.717608439 | −9.622596776 | 28.65893980 |
| 9.0 | −7.494744030 | −8.276958009 | 24.64927991 |
| 10.0 | −8.601421904 | −7.222980039 | 28.83748475 |

**Table 3.**
*Numerical solution of the Lorenz system using RK4 method at step size 0.1.*

**Solution**
The number of iteration needed is determine thus:

$$t_n = t_0 + nh$$

$$n = \frac{t_n - t_0}{h} = \frac{10 - 0}{0.1} = 100$$

The numerical solution is generated using Maple mathematical programming software. The codes and solution are presented below (**Table 3**).

**Algorithm/Maple programming Code of the numerical solution of Lorenz system using RK4 method.**

```
## This is the numerical solution of Lorenz system using RK 4method##
  sys1:=diff(x(t),t)=sigma*(y(t)-x(t)),diff(y(t),t)=-x(t)*z(t)+r*x(t)-y(t),diff(z
  (t),t)=x(t)*y(t)-b*z(t);
  ##
  restart:
X:=array(0..20000):Y:=array(0..20000):Z:=array(0..20000):T:=array(0..20000):A:
=0.0:h:=0.1:N:=100:sigma:=10.0:b:=(8/3):r:=28.0:
  for m from 0 to N do T[m]:=A+m*h:X[0]:=0.1:Y[0]:=0.1:Z[0]:=0.1:od:
  for m from 0 to N-1 do K1:=h*(sigma*(Y[m]-X[m])): M1:=h*(-X[m]*Z
  [m]+r*X[m]-Y[m]):N1:=h*(X[m]*Y[m]-b*Z[m]):K2:=h*(sigma*((Y[m]+
  M1/2)-(X[m]+K1/2))): M2:=h*(-(X[m]+K1/2)*(Z[m]+N1/2)+r*(X[m]+
  K1/2)-(Y[m]+M1/2)):N2:=h*((X[m]+K1/2)*(Y[m]+M1/2)-b*(Z[m]+N1/2))
  :K3:=h*(sigma*((Y[m]+M2/2)-(X[m]+K2/2))): M3:=h*(-(X[m]+K2/2)*(Z
  [m]+N2/2)+r*(X[m]+K2/2)-(Y[m]+M2/2)):N3:=h*((X[m]+K2/2)*(Y[m]+
  M2/2)-b*(Z[m]+N2/2)):K4:=h*(sigma*((Y[m]+M3)-(X[m]+K3))):M4:=h*
  (-(X[m]+K3)*(Z[m]+N3)+r*(X[m]+K3)-(Y[m]+M3)):N4:=h*((X[m]+K3)*(Y
  [m]+M3)-b*(Z[m]+N3)):X[m+1]:=X[m]+(K1+2*K2+2*K3+K4)/6:Y[m+1]:=Y
  [m]+(M1+2*M2+2*M3+M4)/6: Z[m+1]:=Z[m]+(N1+2*N2+2*N3+N4)/6:od:
  for m from 0 by 10 to N do print (m, T[m], X[m],Y[m],Z[m]);od:
```

For further insight on programming with Maple see [6].

## 4. Conclusions

This chapter discusses the numerical solution of differential equation using Euler and Runge-Kutta methods. The formulas were derived and illustrations given to understand their applications. Algorithms written in Maple computational environment were provided for better understanding and further practice.

## Author details

Victor Akinsola
Computational Laboratory, Department of Mathematics, Adeleke University, Ede, Nigeria

*Address all correspondence to: akinsolaolajide@adelekeuniversity.edu.ng; solajide123@gmail.com

## References

[1] Akinsola VO, Oluyo TO. Mathematical analysis with numerical solutions of the mathematical model for the complications and control of diabetes mellitus. Journal of Statistics and Management Systems. 2019;**22**(5): 845-869. DOI: 10.1080/09720510.2018.1556409

[2] Bronshtein IN, Semendyayev KA, Musiol G, Mühlig H. Handbook of Mathematics. 6th ed. London: Springer; 2015

[3] Baumann G. Mathematics for Engineers IV: Numerics. München: Oldenbourg Wissenschaftsverlag; 2010

[4] Lynch S. Dynamical Systems with Applications using Maple. 2nd ed. Basel: Birkhäuser Verlag AG, Springer; 2010

[5] Shah NH. Numerical Methods with C + + Programming. New Delhi: PHI Learning; 2009. p. 243

[6] Richard H. Computer Algebra Recipes for Mathematical Physics. Boston: Birkhauser; 2005

[7] Montoya CA, Sánchez RD, Castaño LF. Approach to the numerical solution of Lorenz system on SoC FPGA. In: 2016 XXI Symposium on Signal Processing, Images and Artificial Vision (STSIVA). Colombia: Universidad Pontificia Bolivariana (UPB) Seccional Bucaramanga; 2016. pp. 1-4. DOI: 10.1109/STSIVA.2016.7743305

[8] Enns RH, McGuire GC. Nonlinear Physics with Maple for Scientists and Engineers. 2nd ed. Massachusetts, USA: Birkhäuser Boston, Springer Science +Business Media; 2000

[9] Steele WA, Vallauri R. Computer simulations of pair dynamics in molecular fluids. Molecular Physics. 1987;**61**(4):1019-1030. DOI: 10.1080/00268978700101621

[10] Zou Y. Multi-Variable Calculus, A First Step. Berlin/Boston: Walter de Gruyter GmbH; 2020

Chapter 3

# An Efficient Region Merging Algorithm in Raster Space

*Borut Žalik, David Podgorelec, Niko Lukač, Krista Rizman Žalik and Domen Mongus*


## Abstract

This work introduces a new region merging algorithm operating in raster space represented by a 4-connected graph. Necessary definitions are introduced first to derive a new merging function formally. An implementation is described after that, which consists of two steps: a determination of the shared trails of the input cycles, and construction of the resulting merged region. The cycles defining the regions are represented by the Freeman crack chain code in four directions. The algorithm works in linear time $O(n)$, where $n$ is the number of total graph vertices, i.e. pixels. However, the expected time complexity for one merging operation performed by the algorithm is $O(1)$.


**Keywords:** computer science, algorithms, 4-connected graph, merging function, chain code

## 1. Introduction

Region merging is one of the most commonly performed tasks in image processing that enables Object-Based Image Analysis (OBIA). Early approaches to OBIA performed image segmentation by the classical split and merge approach. Here, a meaningful partition was defined by applying a split process to define a set of elementary (homogeneous) regions that are then merged under certain conditions [1]. The latter may be based on geometric attributes like area, texture attributes like statistical moments of intensity distribution, shape attributes like shape factors, or any of their combinations [2–4]. On the other hand, more recent approaches to OBIA focus on hierarchical image segmentations that are based on scale-space representation, i.e. a set of image segmentations at different detail levels, in which the segmentation at finer levels are nested with respect to those at coarser levels [1, 5]. Some popular examples of such hierarchies include max-tree [6], $\alpha$-tree [7, 8], and watershed hierarchies [9]. Unfortunately, hierarchical segmentation results in a huge number of nested partitions, which have to be merged efficiently. The region merging becomes in this way one of the most critical parts of the segmentation process.

Region merging can be considered from different theoretical aspects. A set merging problem, which has a long history in computing, is the first of them. Hopcroft and Ullman [10] proposed two algorithms based on quadtrees, both working in $O(n \log n)$ time, where $n$ is the number of elements in the sets. In the first algorithm, the elements

 IntechOpen

can be placed only in the leaves, while, in the second algorithm, the elements can exist in any of the tree vertices. Another tree-based algorithm was proposed by Tarjan [11] with the time complexity of $O(m\ \alpha(m,n))$, where $\alpha(m,n)$ is related with the inverse Ackermann function, while $m$ and $n$ correspond to the numbers of elements in both sets. His algorithm was also used by Najman et al. [9] for hierarchical watershed cuts. Tarjan and van Leeuwen [12] performed the worst-case analysis of the algorithms and concluded that the linear-time set-merging algorithm remains an open problem. Cormen et al. [13] also considered merging of the disjoint sets using either linked lists or trees. Another solution for merging regions was introduced by Horowitz and Pavlidis [14]. This method is also based on quadtrees, with, as pointed out by Brun and Domenger, considerable limitations [15]. They recognised that the regions differ importantly from the classical understanding of the sets. Namely, the elements of the regions also have spatial attributes (i.e. raster coordinates), and, therefore, it is possible to determine the border of the regions uniquely. Brun and Domenger developed a method by placing the image in the Khalimsky plane [16]. The region is considered as a set of topological maps which are mapped in the Euclidean plane. Another approach is based on the theory of geometric and solid modelling [17, 18], where merging is considered as a special case of the Boolean union. The so-called regularised Boolean operations were introduced to preserve the dimension homogeneity of the resulting object [19]. The solution is, typically, found in two steps. First, the intersection points between the involved geometric objects are determined, and second, the resulting shape is determined by the so-called walkabout. In 2D, the first part is solved in the expected time $O((n+m)\log(n+m)+I)$, where $n$ and $m$ are the number of vertices determining the input polygons, and $I$ is the number of actual intersections [20]. If the proper data structure is used, the second step is realised in linear time. Such data structures have been proposed by Grainer and Horman [21], Vatti [22], and Liu et al. [23]. Rivero and Feito [24] proposed an approach for Boolean operations on polygons based on the theory of simplices. Their idea was later improved in ref. [25]. Very recently, an algorithm for Boolean operations for rasterised shapes was presented in ref. [26]. A space-filling curve was applied for the determination of the intersected pixels, while the walkabout was performed with a Greiner and Horman-like data structure. The proposed geometric approaches, however, cannot be applied in the OBIA, as they are based on the theory of regularised Boolean operations, which preserves the dimensional homogeneity of the resulting objects strictly. Consequently, this approach cannot handle all possible cases which may appear during region growth.

In this chapter, a new solution is proposed for a general region merging problem suitable for hierarchical OBIA. The main contributions are a theoretical derivation of the merging function in the raster space, represented by a 4-connected graph, and a proposal of an efficient implementation based on chain codes that ensure compact region representation.

The chapter is structured in five sections. Section 2 introduces the problem and formalises it. Brief implementation hints are given in Section 3. Section 4 presents empirical results, while Section 5 concludes the chapter.

## 2. Definitions

The key terms, needed to present the problem and to derive its formal solution, are defined in this section. Among other concepts, the region, raster space, and region merging are defined, which appeared in the title of this chapter.

**Directed graph.** $G = (V, E)$ defined by a vertex set $V = \{v_i\}$ and an edge set $E = \{e_{i,j}\}$ is a directed graph if $E$ is given by ordered pairs (directed edges) of vertices $e_{i,j} = (v_i, v_j)$.

**Raster space.** Let $G = (V, E)$ be a directed graph. If $V$ is determined by regularly spaced vertices $v_i = (x_i, y_i)$ with integer coordinates $x_i \in [0, X]$ and $y_i \in [0, Y]$, and $E$ imposes 4-connectivity on them, then $G$ defines the raster space. In other words, for each pair of adjacent vertices $v_i, v_j$ linked by edge $e_{i,j} \in E$, there exists either relation $e_{i,j} \to (x_j, y_j) = (x_i \pm 1, y_i)$ or $e_{i,j} \to (x_j, y_j) = (x_i, y_i \pm 1)$. Each vertex can also be linked to itself, thus, $\forall v_i \in V \to e_{i,i} \in E$.

Intuitively, a region is a group of connected raster cells (grid cells or pixels). It may be represented either as a collection of the pixels themselves or by its boundary. This second possibility is used in this work. It is based on the concepts of trail and cycle which must, therefore, be introduced first.

**Trail.** Trail $t_{i_0,i_L} = \langle v_{i_0}, v_{i_1}, \dots, v_{i_L} \rangle$ in G with length $L$ is a sequence of adjacent vertices where for each pair $v_{i_l}, v_{i_{l+1}} \in t_{i_0,i_L} \to e_{i_l,i_{l+1}} \in E$. Trails $t_{i_0,i_L}$ and $t_{j_0,j_K}$ are connected if they share at least one subtrail, i.e. if a set of subtrails $T = t_{i_0,i_L} \cap t_{j_0,j_K} \neq \emptyset$.

**Figure 1** shows two cases of connected trails. Trails $t_{0,6}$ and $t_{7,8}$ are connected through subtrail $t_{4,4} = \langle v_4, v_4 \rangle$ in **Figure 1a**, while trails $t_{0,6}$ and $t_{9,7}$ in **Figure 1b** share two subtrails, namely $T = t_{0,6} \cap t_{9,7} = \{t_{1,1}, t_{3,5}\}$, where $t_{1,1} = \langle v_1, v_1 \rangle$ and $t_{3,5} = \langle v_3, v_4, v_5 \rangle$. The shared subtrails will be hereinafter referred to as the intersection trails.

Trail $t_{i_0,i_L}$ can be split into two trails $t_{i_0,i_l}$ and $t_{i_{l+1},i_L}$ at any $v_{i_l} \in t_{i_0,i_L}$. $t_{i_0,i_L}$ is, therefore, a concatenation of $t_{i_0,i_l}$ and $t_{i_{l+1},i_L}$ as formally shown in Eq. (1):

$$t_{i_0,i_L} = t_{i_0,i_l} \frown t_{i_{l+1},i_L}. \tag{1}$$

**Cycle.** Trail $t_{i_0,i_L} = \langle v_{i_0}, v_{i_1}, \dots, v_{i_{L+1}} \rangle$ is cycle $c_{i_0,i_L} = \langle v_{i_0}, v_{i_1}, \dots, v_{i_L} \rangle$, if $i_0 = i_{L+1}$. As each vertex can be linked to itself, the smallest cycle $c_{i_0,i_0} = \langle v_{i_0} \rangle$ is defined by trail $t_{i_0,i_0} = \langle v_{i_0}, v_{i_0} \rangle$. Contrary to the traditional definition of cycle, we do require that all vertices, except the end vertices, are distinct in $c_{i_0,i_L}$. Any cycle can, because of this, be composed of more than one cycle, where intermediate vertices are contained more than once.

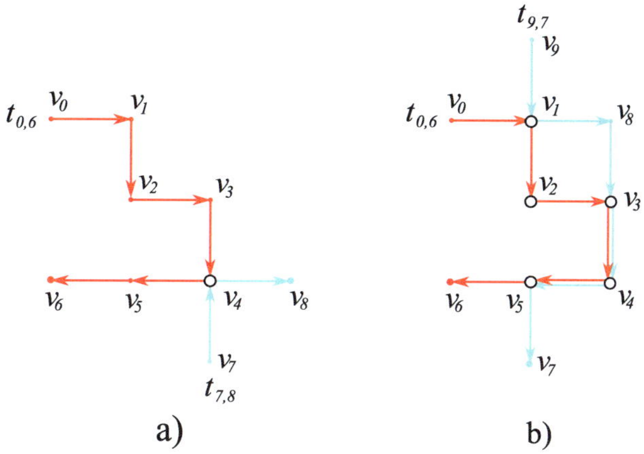


**Figure 1.**
*Connected trails: (a)* $t_{0,6} = \langle v_0, v_1, v_2, v_3, v_4, v_5, v_6 \rangle$, $t_{7,8} = \langle v_7, v_4, v_8 \rangle$, *(b)* $t_{0,6} = \langle v_0, v_1, v_2, v_3, v_4, v_5, v_6 \rangle$, $t_{9,7} = \langle v_9, v_1, v_8, v_3, v_4, v_5, v_7 \rangle$.

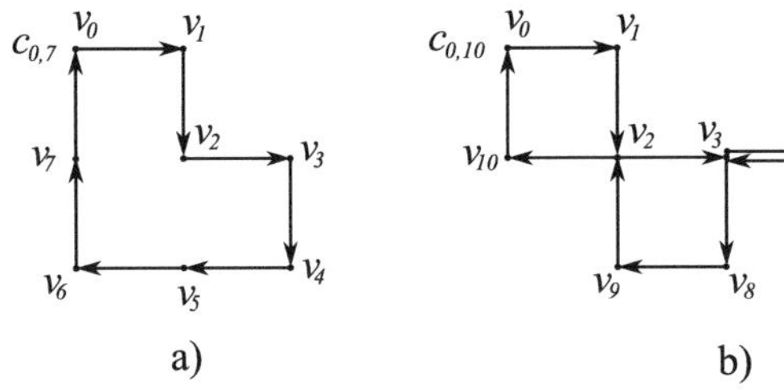


**Figure 2.**
*Cycles: (a)* $c_{0,7} = \langle v_0, v_1, v_2, v_3, v_4, v_5, v_6, v_7 \rangle$*, (b)* $c_{0,10} = \langle v_0, v_1, v_2, v_3, v_4, v_5, v_6, v_7, v_4, v_3, v_8, v_9, v_2, v_{10} \rangle$.

**Figure 2** shows examples of two cycles. The one in **Figure 2a** contains each vertex exactly once, while vertices $v_2$, $v_3$, and $v_4$ are contained twice in the cycle in **Figure 2b**. Note that any cycle can be rotated by any number of vertices $0 < l \leq L$, i.e. $c_{i_0,i_L} = \langle v_{i_0}, v_{i_1}, \ldots, v_{i_L} \rangle = \langle v_{i_l}, v_{i_{l+1}}, \ldots, v_{i_L}, v_{i_0}, v_{i_1}, \ldots, v_{i_{l-1}} \rangle = c_{i_l,i_{l-1}}$. Its decomposition can then be described as concatenation from Eq. (2):

$$c_{i_0,i_L} = t_{i_l,i_k} {}^\frown t_{i_{k+1},l_{L-1}}. \tag{2}$$

As shown in **Figure 3**, any subtrail $t_{i_l,i_k} \subseteq c_{i_0,i_L}$, $0 \leq l, k \leq L$, can be removed from cycle $c_{i_0,i_L}$ according to Eq.(3). The obtained result is also a subtrail.

$$t_{i_{k+1},i_{l-1}} = c_{i_0,i_L} \backslash t_{i_l,i_k}. \tag{3}$$

Let $c_{i_0,i_3} = \langle v_{i_0}, v_{i_1}, v_{i_2}, v_{i_3} \rangle$ be an elementary clockwise oriented cycle, where $v_{i_0} = (x_i, y_i), v_{i_1} = (x_i, y_i + 1), v_{i_2} = (x_i + 1, y_i + 1)$, and $v_{i_3} = (x_i + 1, y_i)$. This elementary cycle defines a grid cell, with its interior on the right side of each edge $e_{i_l,i_{l+1}} = (v_{i_l}, v_{i_{l+1}})$, $0 \leq l \leq 3$ as shown in **Figure 4**.

**Region.** The region $R$ is either a grid cell defined by an elementary clockwise oriented cycle or a group of grid cells bounded by the resulting cycle(s) of the region merging function (defined soon after this definition). For simplicity, a region will be equated with its boundary in the continuation, i.e. $R$ will be treated as a cycle or a set of cycles.

**Figure 5** shows the result of merging two elementary cycles $c_{i_0,i_L}$ and $c_{j_0,j_K}$ $(L = K = 3)$, which share either an edge (**Figure 5a**) or a vertex (**Figure 5b**). It indicates that the resulting merged region is defined by a concatenation of both

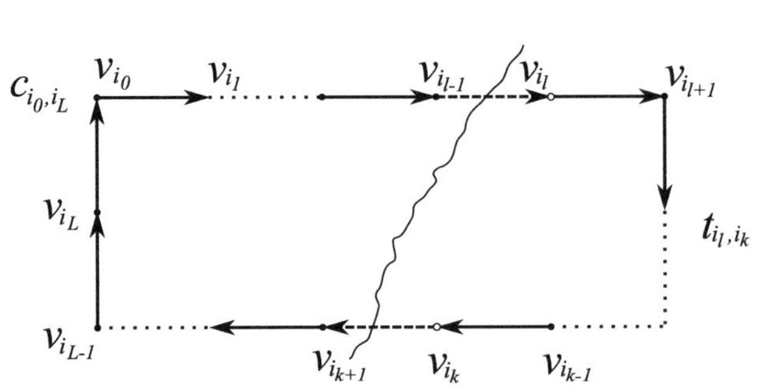

**Figure 3.**
*Removing trail* $t_{i_l,i_k}$ *(on the right) from cycle* $c_{i_0,i_L}$ *results in subtrail* $t_{i_{k+1},i_{l-1}}$ *(on the left).*

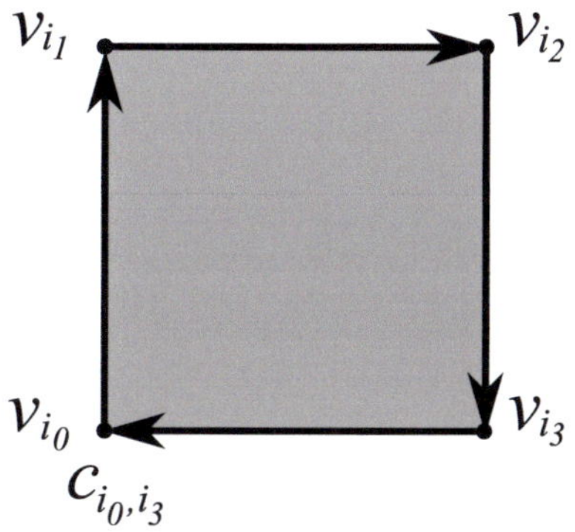


**Figure 4.**
*Elementary cycle $c_{i_0,i_3}$ enclosing a grid cell.*

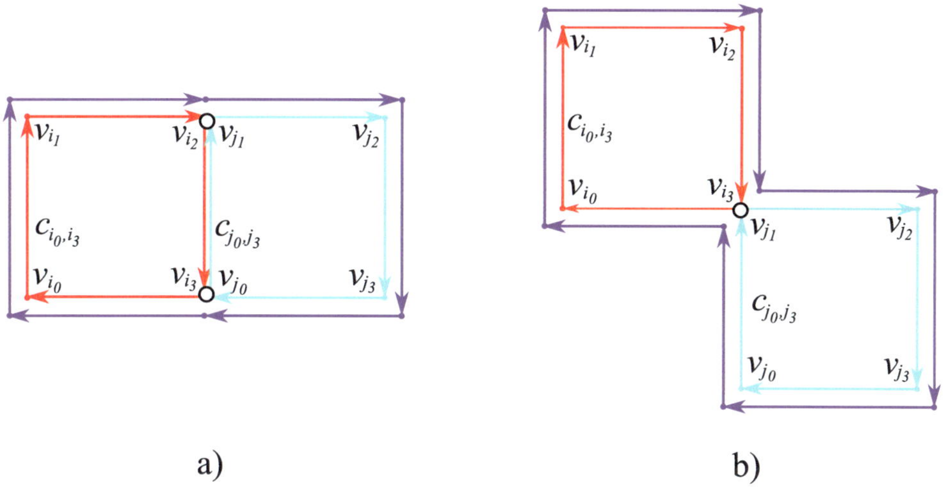


**Figure 5.**
*Merging two elementary cycles with a shared edge (a) and vertex (b); the resulting cycles are in violet.*

elementary cycles without their intersection $c_{i_0,i_L} \cap c_{j_0,j_K}$. Note, however, that $c_{i_0,i_L}$ and $c_{j_0,j_K}$ are both oriented in the clockwise direction, and, therefore, the orientation of the shared edges is opposite. To make them equal and, thus, to make their intersection non-empty, the orientation changing operation $\leftarrow$ is defined here, such that $\overleftarrow{c_{i_0,i_L}} = \langle v_{i_L}, v_{i_{L-1}}, \dots, v_{i_0} \rangle$. **Figure 6** shows an example of merging two non-elementary cycles which still share a single, but longer intersection trail. A similar conclusion as above may be made. The region merging function may be formally defined now.

**Region merging function.** Two cycles $c_{i_0,i_L}$ and $c_{j_0,j_K}$, which share a single intersection trail $t_{i_l,i_k}$ can be merged into a region $R$ by a merging function $\mathcal{M}$ defined by Eq. (4):

$$\begin{aligned}
\mathcal{M}\left(c_{i_0,i_L}, c_{j_0,j_K}, t_{i_l,i_k}\right) &= \left(c_{i_0,i_L} \setminus \left(c_{i_0,i_L} \cap \overleftarrow{c_{j_0,j_K}}\right)\right)^\frown \langle v_{i_l} \rangle^\frown \left(c_{j_0,j_K} \setminus \left(c_{j_0,j_K} \cap \overleftarrow{c_{i_0,i_L}}\right)\right)^\frown \langle v_{i_k} \rangle = \\
&= \left(c_{i_0,i_L} \setminus t_{i_l,i_k}\right)^\frown \langle v_{i_l} \rangle^\frown \left(c_{j_0,j_K} \setminus t_{j_n,j_m}\right)^\frown \langle v_{i_k} \rangle = \\
&= t_{i_{k+1},i_{l-1}}{}^\frown \langle v_{i_l} \rangle^\frown t_{j_{m+1},j_{n-1}}{}^\frown \langle v_{i_k} \rangle = \\
&= t_{i_{k+1},i_{l-1}}{}^\frown \langle v_{j_m} \rangle^\frown t_{j_{m+1},j_{n-1}}{}^\frown \langle v_{j_n} \rangle.
\end{aligned} \tag{4}$$

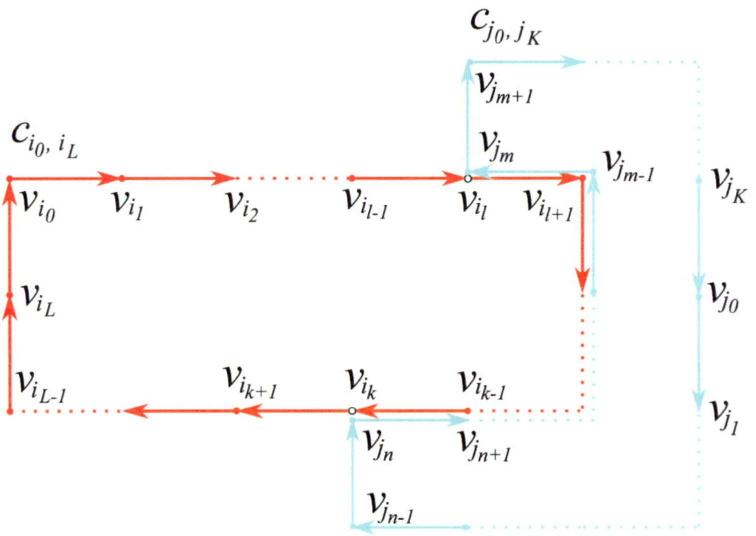


**Figure 6.**
*Merging of two non-elementary cycles whose intersection trail is a longer sequence of edges.*

In a general case, where the intersection of the input cycles consists of more trails, the merged region R is defined by Eq. (5):

$$R = \bigcap_{t^{(h)}_{i_l,i_k} \in T_i} \mathcal{M}\left(c_{i_0,i_L}, c_{j_0,j_K}.t^{(h)}_{i_l,i_k}\right). \tag{5}$$

Eq. (4) is thus applied when $|T_i| = |T_j| = 1$, where $T_i = \{t_{i_k,i_l}\} = c_{i_0,i_L} \cap \overleftarrow{c_{j_0,j_K}}$ and $T_j = \{t_{j_m,j_n}\} = c_{j_0,j_K} \cap \overleftarrow{c_{i_0,i_L}}$. On the other hand $|T_i| = |T_j| > 1$ implies utilisation of

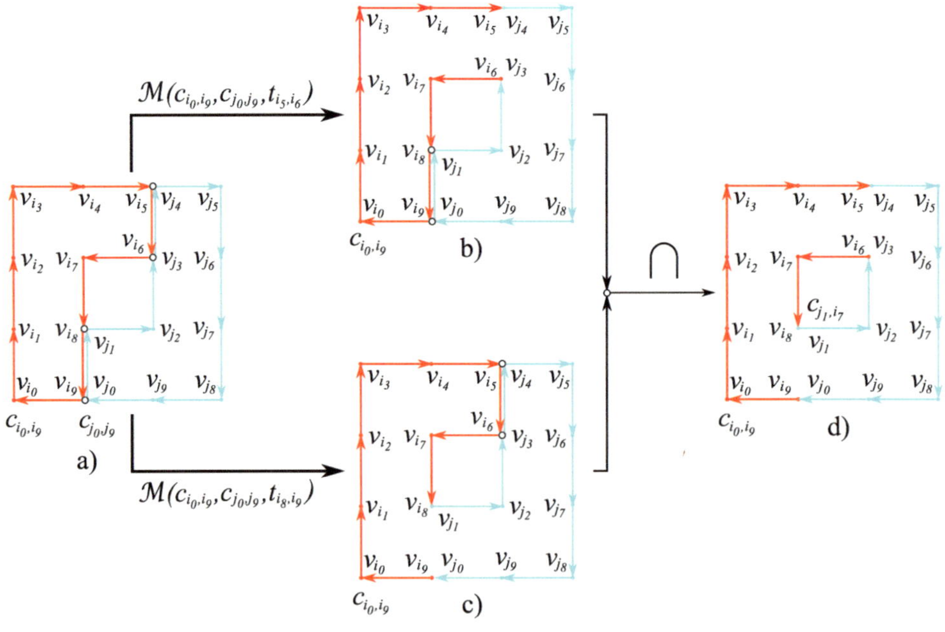


**Figure 7.**
*Applying Eq. (5) to merge two cycles whose intersection consists of two intersection trails, i.e.* $|T_i| = 2$.

Eq. (5) and each intersection trail then results in a new cycle. $|T_i|$ cycles are, therefore, constructed, and the resulting region is described by the intersection of these cycles. Note that $h$, such that $0 \leq h < |T_i|$, is an index of the intersection trail in Eq. (5). It is obvious that Eq. (5) is valid also when $|T_i| = 1$.

**Figure 7** shows an illustrative example. The set of intersection trails is $T_i = \{t_{i_5,i_6}, t_{i_8,i_9}\}$. Applying Eq. (4) on the intersection trail $t_{i_5,i_6} \in T_i$ results in **Figure 7b**. Similarly, using Eq. (4) on the second intersection trail $t_{i_8,i_9} \in T_i$ gives **Figure 7c**. The final result is then obtained as an intersection (Eq. (5)) between cycles from **Figure 7b** and **c**. As seen in **Figure 7d**, the resulting region $R$ consists of two cycles, which are exactly the same as $|T_i|$.

## 3. Implementation

The concept of chain codes is used to represent regions $R \subseteq G$ in the presented method. The chain code, introduced by Freeman [27], consists of a few simple commands by which navigation through the edges of $G$ is made possible. Freeman proposed two chain codes known as Freeman chain code in eight ($F8$) and four ($F4$) directions. Other chain codes were discovered by Bribiesca (Vertex Chain Code, *VCC*) [28], Sánchez-Cruz and Rodríguez-Dagnino (Three-Orthogonal chain code, $3OT$) [29], Žalik et al. (Unsigned Manhattan Chain Code, *UMCC*) [30], and Dunkelberger and Mitchell (Mid-crack Chain Code) [31]. In general, there are two types of chain codes: those operating on raster pixels and those working with raster edges. The latter is known as crack-chain codes [32, 33]. $F4$ is the only chain code which can be used in both contexts, and its crack interpretation is used in this algorithm.

The $F4$ alphabet consists of four commands/symbols $\sigma_i \in \Sigma_{F4}$, $\Sigma_{F4} = \{0, 1, 2, 3\}$, shown in **Figure 8**. Let $\langle \sigma_i \rangle$ be the sequence of $F4$ commands. To embed the chain code in $G$, the position of the chain code starting vertex $v_0$ is needed, while the positions of the remaining vertices are determined from the $F4$ commands according to Eq. (6):

$$v_{i+1} = \begin{cases} (x_i + 1, y_i), & \text{if } \sigma_i = 0; \\ (x_i, y_i - 1), & \text{if } \sigma_i = 1; \\ (x_i - 1, y_i), & \text{if } \sigma_i = 2; \\ (x_i, y_i + 1), & \text{if } \sigma_i = 3. \end{cases} \tag{6}$$

**Figure 8b** shows the elementary cycle $c_{i_0,i_3}$ determined as $v_{i_0}\langle 1, 0, 3, 2 \rangle$, where $v_{i_0} = \left(x_{i_0}, y_{i_0}\right)$ are the coordinates of the cycle's starting vertex.

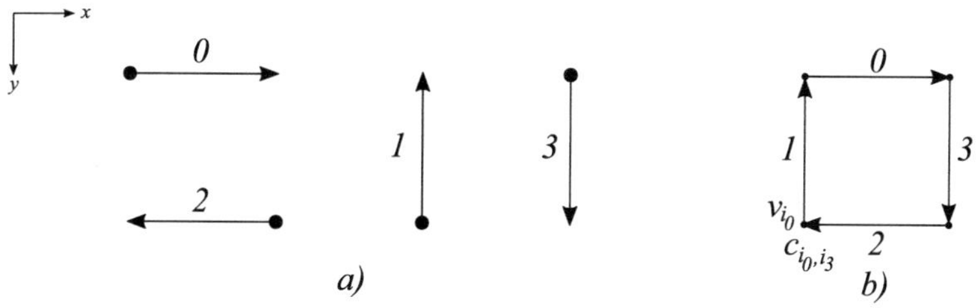


**Figure 8.**
*F4 chain code symbols (a); the elementary cycle described by F4 chain code symbols (b).*

Any region in $G$ can be represented in this way. For example, regions containing more cycles are shown in **Figure 9**, where, for better presentation, the inner cells of the region are shadowed. According to the definitions from Section 2, a vertex $v_i \in R$ can be part of two cycles at the same time, or, within each cycle, $v_i$ can be passed twice, too. In the continuation, the cycle corresponding to the outer border of $R$ is considered as a *loop*, while cycles representing holes are named *rings* [19]. The orientation of the loop is clockwise, while the rings' is the opposite (**Figure 9**).

A data structure for representing $R$ is shown schematically in **Figure 10**. It consists of an array of starting points and an array of $F4$ chain code sequences. The loop is always located at index 0, and $k$, $k \geq 0$, rings follow in an arbitrary order. The algorithm, which implements Eq. (5), consists of two main steps:

- Determining the intersection trails $T_i$ between cycles of input regions and
- Realising Eq. (5) by performing a walkabout through the edges not in $T_i$.

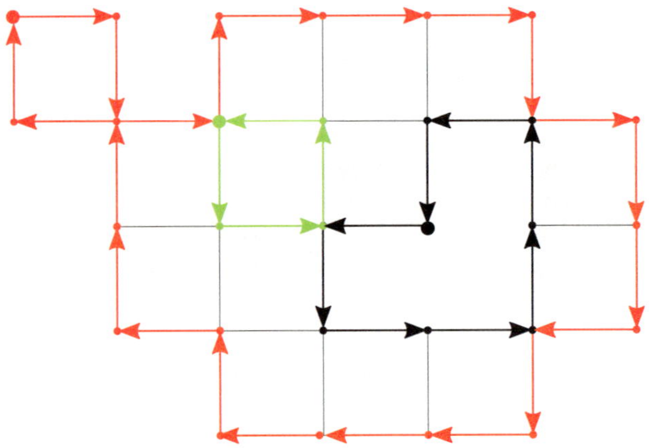

**Figure 9.**
*Region with two rings: The rings' vertices (green, black) can be shared with the loop (red); the vertices within the loop can be used twice.*

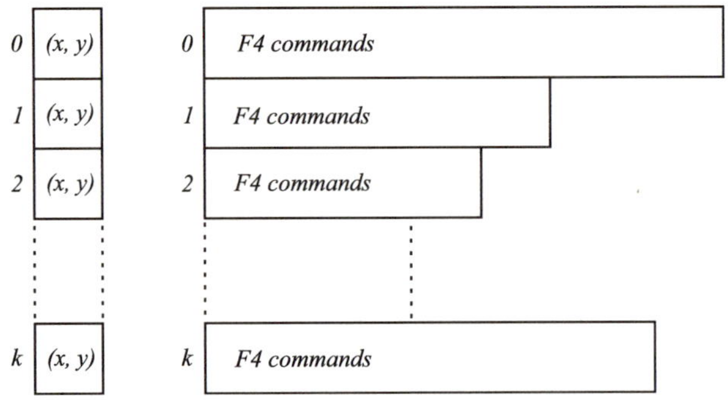


**Figure 10.**
*Data structure of region R: Array of starting points (left) of individual cycles and the corresponding sequences of F4 chain codes (right).*

### 3.1 Determining the intersection trails

Determination of intersection points between edges of regions can be related with the problem of finding intersections between polygon edges. As the naive implementation of the latter works with $O(n^2)$ time complexity, various approaches were suggested to reduce it [13, 34–36]. The presented solution exploits the fact that regions $R_i$ and $R_j$ are embedded into common directed graph $G$, i.e. $R_i \in G \wedge R_j \in G$. The following data are associated with each vertex $v_i \in G$:

- two pointers ($P_{i1}$ and $P_{i2}$) into an array of $F4$ symbols for region $R_i$ (one or both pointers can be NULL),
- two pointers ($LR_{i1}$ and $LR_{i2}$) pointing to the loop, or to the corresponding ring of region $R_i$ (similarly as above, one or both pointers can be NULL), and
- the same information for region $R_j$.

Let us consider the example in **Figure 11**, where the loop's edges of $R_i$ are plotted in red, the edges of its ring in black, while edges of region $R_j$ are in cyan. The content of data structures for both regions is given in **Table 1**. **Table 2** shows the information of some characteristic vertices in $G$. Vertex $v_a$, for example, belongs to $R_i$, its $P_{i1}$ points to the index 1 in the array of $F4$ chain codes, and the vertex belongs to the loop ($LR_{i2} = 0$). Vertex $v_f$ is met two times by the edges of the $R_i$ loop and, therefore, two pointers are pointing to the 3rd and 15th positions. Vertex $v_h$ is the most interesting. In this vertex three cycles are met. Pointer $P_{i1}$ points to the $7^{th}$ $R_i$ loop

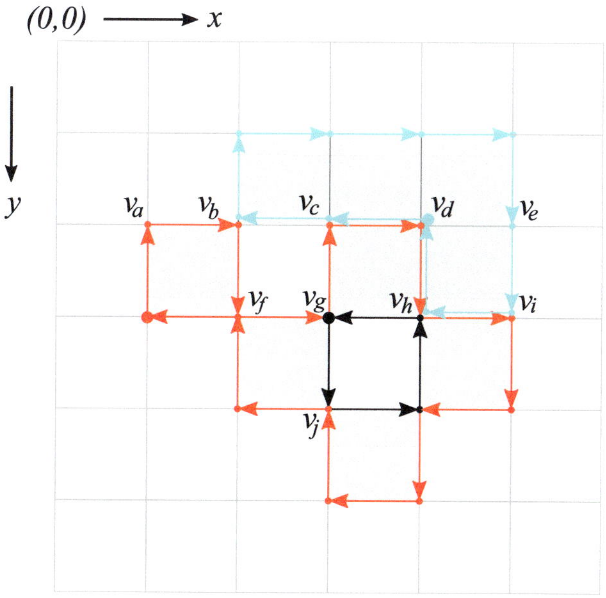


**Figure 11.**
*Regions $R_i$ (red and black) and $R_j$ (cyan) embedded into G, and some characteristic vertices considered in* ***Table 2****.*

| | | | **i:** | **0** | **1** | **2** | **3** | **4** | **5** | **6** | **7** | **8** | **9** | **10** | **11** | **12** | **13** | **14** | **15** |
|---|---|---|---|---|---|---|---|---|---|---|---|---|---|---|---|---|---|---|---|
| $R_i$ | 0 | (1, 3) | $F4$: | 1 | 0 | 3 | 0 | 1 | 0 | 3 | 0 | 3 | 2 | 3 | 2 | 1 | 2 | 1 | 2 |
| | | | i: | 0 | 1 | 2 | 3 | | | | | | | | | | | | |
| | 1 | (3, 3) | $F4$: | 3 | 0 | 1 | 2 | | | | | | | | | | | | |
| $R_j$ | | | i: | 0 | 1 | 2 | 3 | 4 | 5 | 6 | 7 | 8 | 9 | | | | | | |
| | 0 | (4, 2) | $F4$: | 2 | 2 | 1 | 0 | 0 | 0 | 3 | 3 | 2 | 1 | | | | | | |

**Table 1.**
*Data structures for regions $R_i$ and $R_j$ from* ***Figure 11****.*

| **Vertex in Figure 11** | $P_{i1}$ | $P_{i2}$ | $LR_{i1}$ | $LR_{i2}$ | $P_{j1}$ | $P_{j2}$ | $LR_{j1}$ | $LR_{j2}$ |
|---|---|---|---|---|---|---|---|---|
| $v_a$ | 1 | / | 0 | / | / | / | / | / |
| $v_b$ | 2 | / | 0 | / | 2 | / | 0 | / |
| $v_c$ | 5 | / | 0 | / | 1 | / | 0 | / |
| $v_d$ | 6 | / | 0 | / | 0 | / | 0 | / |
| $v_e$ | / | / | / | / | 7 | / | / | / |
| $v_f$ | 3 | 15 | 0 | 0 | / | / | / | / |
| $v_g$ | 4 | 0 | 0 | 1 | / | / | / | / |
| $v_h$ | 7 | 3 | 0 | 1 | 9 | / | 0 | / |
| $v_i$ | 8 | / | 0 | / | 8 | / | 0 | / |
| $v_j$ | 13 | 1 | 0 | 1 | / | / | / | / |
| *(NULL pointers are marked with '/')* | | | | | | | | |

**Table 2.**
*The content of G at specific vertices marked in* ***Figure 11****.*

position, and $P_{i2}$ points to the $3^{rd}$ position of the first $R_i$ ring. The loop of $R_j$ is accessed by the chain code command stored at position 9 in the $F4$ array.

Having marked the vertices in G properly, it is easy to determine the intersection trails. The region with the smallest number of edges is found (let us suppose it is $R_j$), and all its vertices are visited. The sequence of edges marked with pointers of both regions is identified as being a part of the intersection trail.

## 3.2 Performing the walkabout

Those trails which were not labelled as the intersection ones, are united into the new region by the algorithm, consisting of the following steps:

1. Mark all edges from intersection trails as *visited* and the remaining edges as *not visited*.

2. Find an arbitrary non-visited edge $e \in R_j$. If such edge does not exist, jump to step 9.

3. The initial vertex of $e$ is chosen as the starting vertex $v_s$ for the walkabout. An empty queue $Q$ is created.

4. Walk along the edges of $R_j$, mark each passed edge as *visited* and store passed $F4$ commands into $Q$ until $v_s$ is met, or the vertex with set $R_i$ pointer is reached.

5. If $v_s$ is reached, go to step 8, otherwise switch to the edge of region $R_i$ using the pointer stored at the vertex from $G$.

6. Walk along the edges of $R_i$, mark each passed edge as *visited* and store passed $F4$ commands into $Q$ until the $v_s$ is met or the vertex with $R_j$ pointer is detected.

7. If $v_s$ is not reached yet, switch to the region $R_j$ using the pointer stored at the considered vertex, and go to step 4, otherwise go to the next step.

8. Store the obtained cycle into the list of cycles $LoC$ and return to step 2.

9. Insert non-visited cycles of input both regions $R_i$ and $R_j$ into $LoC$.

10. Find the loop in $LoC$; all others cycles in $LoC$ represent rings of the merged region $R$.

These steps are highlighted in Algorithm 1. The decision whether the cycle defines a loop or a ring depends on its orientation. It can be determined by Eq. (7), where $Q = \langle \sigma_i \rangle$ ($F4$ chain code commands $\sigma_i$ are treated as integers for this purpose).

$$o = \sum_{i=0}^{|Q|-1} \begin{cases} -1 & : \sigma_i = 0 \wedge \sigma_{(i+1) mod \ |\Sigma_{F4}|} = |\Sigma_{F4}| - 1 \\ 1 & : \sigma_i = |\Sigma_{F4}| - 1 \wedge \sigma_{(i+1) mod \ |\Sigma_{F4}|} = 0 \\ \sigma_{(i+1) mod \ |\Sigma_{F4}|} - \sigma_i & : \text{otherwise} \end{cases} \quad (7)$$

The equation evaluates each right turn with $-1$ and each left turn with 1. The clockwise oriented cycles result in $o = -4$, while the counter-clockwise cycles achieve $o = 4$.

**Algorithm 1** Merging regions $R_i$ and $R_j$.

**function** MERGE($G$, $R_i$, $R_j$)
  ▷ function merges regions $R_i$ and $R_j$ embedded in graph $G$
  Insert($G$, $R_i$, *NonVisited*) ▷ insert region into $G$, mark edges as *NonVisited*
  Insert($G$, $R_j$, *NonVisited*)
  MarkIntersectionEdges($G$, $R_i$, $R_j$) ▷ intersection edges are marked as *Visited*
  $e$ = GetNonVisitedEdge($R_j$)
  $LoC = \{ \}$ ▷ List of Cycles should be empty
  **while** Visited($R_j$, $e$) = *NonVisited* **do**
    *CycleFound* = **false**
    $Q = \{ \}$ ▷ clear the queue containing F4 symbols
    $Q$ =AddF4Symbol($Q$, $R_j$, $e$) ▷ add F4 chain code symbol of $e$
    $v_s$ = ReturnVertex($R_j$, $e$) ▷ store the starting vertex

```
        MarkVisitedEdge(R_j, e)                                    ▷ mark edge as visited
        e = NextEdge(R_j, e)                 ▷ move one edge forward along R_j boundary
        repeat
          while (Visited(R_j, e) = NonVisited) AND
                (ReturnVertex(R_j, e) ≠ v_s) do               ▷ walk along edges of R_j
            Q = AddF4Symbol(Q, R_j, e)              ▷ add F4 chain code symbol of e
            MarkVisitedEdge(R_j, e)                                ▷ mark edge as visited
            e = NextEdge(R_j, e)             ▷ move one edge forward along R_j boundary
          end while
          if ReturnVertex(R_j, e) = v_s then                       ▷ the cycle is formed
            CycleFound = true
          else                                    ▷ otherwise continue the walk on R_i
            e = ReturnEdgeFromOtherRegion(R_j, e)                          ▷ get e ∈ R_i
            while (Visited(R_i, e) = NonVisited) AND              ▷ walk along R_i edges
                  (ReturnVertex(R_i, e) ≠ v_s) do
              Q = AddF4Symbol(Q, R_i, e)
              MarkVisitedEdge(R_i, e)
              e = NextEdge(R_i, e)
            end while
            if ReturnVertex(R_i, e) = v_s then
              CycleFound = true
            else                                       ▷ jump to the R_j boundary again
              e = ReturnEdgeFromOtherRegion(R_i, e)                        ▷ get e ∈ R_j
            end if
          end if
        until CycleFound = true
        LoC = StoreCycle(v_s, Q)                               ▷ store constructed cycle
        e = GetNonVisitedEdge(R_j)                          ▷ get next non-visited edge
      end while
      AddNonVisitedCycles(LoC, R_i)                  ▷ add non-visited cycles (holes)
      AddNonVisitedCycles(LoC, R_j)
      DetermineLoop(LoC)              ▷ among all cycles the loop is found by (Eq. 7)
      R = ConstructRegion(LoC)
      return R
end function.
```

## 4. Analysis of the algorithm

### 4.1 Time and space complexity estimation

The proposed algorithm consists of three parts: finding the intersection trails, performing the walkabout, and determination of loop and rings.

Let the four-connected graph $G$ consist of $n$ vertices. Let $k_i$ and $k_j$ represent the number of $F4$ edges defining cycles of $R_i$ and $R_j$, respectively. $k_i = k_j = k$ may be assumed without loss of generality. At first, both regions are embedded into $G$. This is done in $T_e(k) = T(k_i) + T(k_j) = 2k$ time. One of the regions is walked-about and the edges of the intersection trails are determined in time $T_w(k) = k$. The first part of the algorithm is, therefore, executed in $T_1(k) = T_e(k) + T_w(k) = 3k$ time.

During the walkabout, all edges of both regions not being part of the intersection trail, are visited exactly once. In the worst case, all $2k$ edges must be visited in time $T_2(k) = 2k$.

The orientation of the cycles of the merged region is determined in the last step. This task is also terminated in $T_3 = 2k$ time, as the merged region cannot have more than $2k$ edges.

The proposed merging algorithm is, therefore, realised in $T(k) = T_1(k) + T_2(k) + T_3(k) = 3k + 2k + 2k = 7k = O(k)$. $k$ cannot be greater than $n$, and therefore, one merging operation is terminated in the worst case in $O(n)$ time. However, in the majority of cases, $k < < n$. In such, an expected case, one merging operation is terminated in a constant time $O(1)$.

The algorithm needs memory for $G$ with $n$ vertices, i.e. $S_G(n) = n$. In addition, two regions need to be stored. In the worst case, both regions require additional $S_R(n) = n$ memory space. The algorithm, therefore, works in $S(n) = 2n = O(n)$ space.

### 4.2 Experiment

The standard benchmark images shown in **Figure 12** have been used in the experiment with different resolutions. A criterion for merging two neighbouring regions was the colour similarity. By doubling the size of the image in both directions iteratively, the number of pixels $n$ is increasing by the power of 4, and the number of required merging operations follows this exponential growth; actually, the number of merging operations is $n - 1$.

**Table 3** shows spent CPU time while performing merging from single pixels up to the entire image. A personal computer with a 3,5 GH Intel® Core™ i5–6600 K processor and 32 G bytes of RAM was used in the experiment. The program was implemented in C++ and compiled with C++ Visual Studio 2019 under the Windows 10 operating system. As can be seen, the actual CPU time spent depends on the colour characteristics of the images. The image Lenna has large parts of very similar colours and therefore, the regions grow rapidly which is reflected in the shortest CPU spent time. The image Peppers exposes a similar characteristic. On the other hand, the image Baboon consists of very small homogeneous regions, reflecting in longer CPU time spent.

a) b) c)

**Figure 12.**
*Images used in the experiments: (a) Lenna, (b) peppers, and (c) baboon.*

| Size (pixels) | $n$ | $t_L$ (s) | $t_P(s)$ | Size (pixels) | $n$ | $t_B(s)$ |
|---|---|---|---|---|---|---|
| 64 × 64 | 4096 | 0.014 | 0.017 | 63 × 60 | 3780 | 0.018 |
| 128 × 128 | 16,384 | 0.081 | 0.064 | 125 × 120 | 15,000 | 0.169 |
| 256 × 256 | 65,536 | 0.715 | 0.624 | 250 × 240 | 60,000 | 1.682 |
| 512 × 512 | 262,144 | 7.885 | 11.778 | 500 × 480 | 240,000 | 20.454 |

**Table 3.**
*Spent CPU time for images Lenna (L), peppers (P), and baboon (B) at different resolutions.*

| Size (pixels) | $t_L$ (s) | $t_P$ (s) | Size (pixels) | $t_B$ (s) |
|---|---|---|---|---|
| 64 × 64 | 0.683 | 0.602 | 63 × 60 | 0.892 |
| 128 × 128 | 8.765 | 8.646 | 125 × 120 | 13.823 |
| 256 × 256 | 219.450 | 211.762 | 250 × 240 | 247.323 |
| 512 × 512 | OOM[a] | OOM[a] | 500 × 480 | OOM[a] |

[a]*OOM denotes out-of-memory.*

**Table 4.**
*Spent CPU time with the referenced approach.*

We implemented the set-based version of the merging operation for comparison. In this case, the region is represented by the STD structure *unordered_map*. The obtained results are shown in **Table 4**. As can be seen, the set-based approach performs considerably slower. In addition, it obviously spends more memory, as images with the highest resolutions cannot be processed any more.

## 5. Conclusion

A new region merging algorithm, suitable for hierarchical object-based image analysis, is proposed in this chapter. The raster space is represented by a 4-connected graph, and a merging function is derived formally upon it. The implementation follows the theoretical investigation strictly. The edges forming the border of the region embedded in the 4-connected graph are represented by the Freeman crack chain code in four directions. The implementation works in two main steps: a determination of the common vertices and edges of the regions being merged, and a walkabout, which realises the theoretically derived merging function. A classification of the obtained region's edges to those representing the holes and those defining the outer border, may be done at the end.

The algorithm's worst-case time complexity is $O(n)$ for one merging operation, where $n$ is the number of graph vertices. However, as the number of edges defining the two regions being merged is much smaller than $n$, the expected time complexity is actually independent on $n$, i.e. the expected time complexity of the proposed algorithm is $O(1)$.

In addition to the methodical implementation of the region merging procedure, the proposed chain code-based approach enables efficient extraction of various essential shape descriptors [3]. The approaches for extracting these descriptors can be divided roughly into the region- and contour-based approaches, and the latter are known as being computationally demanding for traditional hierarchical segmentation and region growing. Namely, they require the boundary to be extracted after each region merging operation. Because of this, these are rarely used, e.g. as stopping criteria during the region growing, or as thresholds for hierarchical cuts. On the other hand, chain codes by themselves allow for efficient description of shapes.

## Acknowledgements

The authors acknowledge the financial support from the Slovenian Research Agency (Research Core Funding No. P2-0041 and Project No. N2-0181), and the Company IGEA, d.o.o., for co-financing this research.

## Thanks

Thanks to Anže Ferčec, who implemented the referenced merging algorithm.

## Nomenclature

| | |
|---|---|
| $c$ | cycle |
| CPU | central processing unit |
| $e$ | edge |
| $E$ | set of edges |
| $F4$ | Freeman chain code in four directions |
| $G$ | directed graph |
| $L$ | the number of vertices in a trail or in a cycle |
| $LR$ | pointer to the loop/ring of the region |
| $LoC$ | list of cycles |
| $\mathcal{M}$ | merging function |
| OBIA | Object-Based Image Analysis |
| $P$ | pointer to the region |
| $Q$ | queue |
| $R$ | region |
| $\Sigma_{F4}$ | alphabet of Freeman's chain code in four directions |
| $\sigma$ | $F4$ symbol |
| $t$ | trail |
| $T$ | set of trails |
| $v$ | vertex |
| $V$ | set of vertices |

## Author details

Borut Žalik*†, David Podgorelec†, Niko Lukač†, Krista Rizman Žalik†
and Domen Mongus†
Faculty of Electrical Engineering and Computer Science, University of Maribor,
Maribor, Slovenia

*Address all correspondence to: borut.zalik@um.si

† These authors contributed equally.

## References

[1] Guimarães SJF, Cousty J, Kenmochi Y, Najman L. A Hierarchical Image segmentation algorithm based on an observation scale. In: Gimel'farb G, Hancock E, Imiya A, Kuijper A, Kudo M, Omachi S, Windeatt T, Yamada K, editors. Structural, Syntactic, and Statistical Pattern Recognition. Lecture Notes in Computer Science;7626. Berlin, Heidelberg: Springer; 2012. pp. 116-125. DOI: 10.1007/978-3-642-34166-3_13

[2] Materka A, Strzelecki M. Texture analysis methods – a review. Lodz, Poland: COST B11 report, Institute of Electronics, Technical University of Lodz; 1998. Available from: https://www.researchgate.net/publication/249723259_Texture_Analysis_Methods_-_A_Review [Accessed: 2021-12-22]

[3] Zhang D, Lu G. Review of shape representation and description techniques. Pattern Recognition. 2004; **37**(1):1-19. DOI: 10.1016/j.patcog.2003.07.008

[4] Kurnianggoro L, Jo KH. A survey of 2D shape representation: Methods, evaluations, and future research directions. Neurocomputing. 2018;**300**: 1-16. DOI: 10.1016/j.neucom.2018.02.093

[5] Xu Y, Carlinet E, Géraud T, Najman L. Efficient computation of attributes and saliency maps on tree-based image representations. In: Benediktsson J, Chanussot J, Najman L, Talbot H, editors. Mathematical Morphology and Its Applications to Signal and Image Processing. Lecture Notes in Computer Science; 9082. Berlin, Heidelberg: Springer; 2015. pp. 693-704. DOI: 10.1007/978-3-319-18720-4_58

[6] Salembier P, Oliveras A, Garrido L. Antiextensive connected operators for image and sequence processing. IEEE Transactions on Image Processing. 1998;**7**(4):555-570. DOI: 10.1109/83.663500

[7] Ouzounis GK, Soille P. Pattern spectra from partition pyramids and hierarchies. In: Soille P, Pesaresi M, Ouzounis GK, editors. Mathematical Morphology and Its Applications to Image and Signal Processing. Berlin, Heidelberg: Lecture Notes in Computer Science; 6671, Springer; 2011. pp. 108-119. DOI: 10.1007/978-3-642-21569-8_10

[8] Ouzounis GK, Soille P. The alpha-tree algorithm: Theory, algorithms and applications. Luxembourg: Technical reports JCR74511, European Commission, Joint Research Centre; 2012. DOI: 10.2788/48773

[9] Najman L, Cousty J, Perret B. Playing with Kruskal: Algorithms for morphological trees in edge-weighted graphs. In: Hendriks CLL, Borgefors G, Strand R, editors. Mathematical Morphology and Its Applications to Signal and Image Processing. Lecture Notes in Computer Science; 7883. Berlin, Heidelberg: Springer; 2013. pp. 135-146. DOI: 10.1007/978-3-642-38294-9_12

[10] Hopcroft JE, Ullman JD. Set merging algorithms. SIAM Journal of Computing. 1973;**2**(4):294-303. DOI: 10.1137/0202024

[11] Tarjan RE. Efficiency of a good but non-linear set union algorithm. Journal of ACM. 1975;**22**(2):215-225. DOI: 10.1145/321879.321884

[12] Tarjan RE, Leeuwen J. Worst-case analysis of set union algorithms. Journal of ACM. 1984;**31**(2):245-281. DOI: 10.1145/62.2160

[13] Cormen TH, Leiserson CE, Rivest RL, Stein C. Introduction to algorithms. 3rd ed. Cambridge, London: MIT; 2009. p. 1292

[14] Horowitz SL, Pavlidis T. Picture segmentation by a tree traversal algorithm. Journal od ACM. 1976; **23**(2):368-388. DOI: 10.1145/321941.321956

[15] Brun L, Domenger JP. A new split and merge algorithm with topological maps. In: Proceedings of the 5th International Conference in Central Europe on Computer Graphics and Visualization (WSCG'97), 10–14 February 1996, Plzen, Czech Republic: University of West Bohemia. p. 21–31. Available from: https://www.researchgate.net/publication/2658919_A_New_Split_and_Merge_Algorithm_with_Topological_Maps [Accessed: 2021-12-22]

[16] Khalimsky E, Kopperman R, Meyer PR. Boundaries in digital planes. Journal of Applied Mathematics and Stochastic Analysis. 1990;**3**(1):27-55. DOI: 10.1155/S1048953390000041

[17] Hoffmann CM. Geometric and solid modeling: An introduction. San Francisco: Morgan Kaufmann; 1989. 338 p. Available from: https://dl.acm.org/doi/book/10.5555/74803 [Accessed: 2021-12-23]

[18] Mäntylä M. An introduction to solid modeling. New York: Computer Science Press; 1987. 401 p. Available from: https://dl.acm.org/doi/book/10.5555/39278 [Accessed: 2021-12-23]

[19] Mortenson ME. Geometric Modeling. 2nd ed. New York: Wiley; 1997. 523 p. Available from: https://dl.acm.org/doi/book/10.5555/248381 [Accessed: 2021-12-23]

[20] Mairson HG, Stolfi J. Reporting and counting intersections between two sets of line segments. In: Earnshaw R, editor. Theoretical Foundations of Computer Graphics and CAD. NATO ASI Series; F40. Berlin, Heidelberg: Springer; 1988. pp. 307-325. DOI: 10.1007/978-3-642-83539-1_11

[21] Greiner G, Hormann K. Efficient clipping of arbitrary polygons. ACM Transaction on Graphics. 1998;**17**(2):71-83. DOI: 10.1145/274363.274364

[22] Vatti BR. A generic solution to polygon clipping. Communications of ACM. 1992;**35**(7):56-63. DOI: 10.1145/129902.129906

[23] Liu YK, Wang XQ, Bao SZ, Gomboši M, Žalik B. An algorithm for polygon clipping, and for determining polygon intersections and unions. Computers & Geosciences. 2007;**33**(5):589-598. DOI: 10.1016/j.cageo.2006.08.008

[24] Rivero M, Feito FR. Boolean operations on general planar polygons. Computers & Graphics. 2000;**24**(6):881-896. DOI: 10.1016/S0097-8493(00)00090

[25] Peng Y, Yong JH, Dong WM, Zhang H, Sun JG. A new algorithm for Boolean operations on general polygons. Computers & Graphics. 2005;**29**(1):57-70. DOI: 10.1016/j.cag.2004.11.001

[26] Žalik B, Mongus D, Rizman Žalik K, Lukač N. Boolean operations on rasterized shapes represented by chain codes using space filling curves. Journal of Visual Communication and Image Representation 2017;49:420–432. DOI: 10.1016/j.jvcir.2017.10.003

[27] Freeman H. On the encoding of arbitrary geometric configurations. IRE Transactions on Electronic Computers. 1961;**EC10**(2):260-268. DOI: 10.1109/TEC.1961.5219197

[28] Bribiesca E. A new chain code. Pattern Recognition. 1999;**32**(2): 235-251. DOI: 10.1016/S0031-3203(98)00132-0

[29] Sánchez-Cruz H, Rodríguez-Dagnino RM. Compressing bi-level images by means of a 3-bit chain code. Optical Engineering. 2005;**44**(9): 097004. DOI: 10.1117/1.2052793

[30] Žalik B, Mongus D, Liu YK, Lukač N. Unsigned Manhattan chain code. Journal of Visual Communication and Image Representation 2016;**38**:186–194. DOI: 10.1016/j.jvcir.2016.03.001

[31] Dunkelberger KA, Mitchell OR. Contour tracing for precision measurements. St. Luis, USA: IEEE International conference on robotics and automation (ICRA); 25–28 March 1985. pp. 22-27. DOI: 10.1109/ROBOT.1985.1087356

[32] Wilson GR. Properties of contour codes. IEE Proceedins – Vision, Image and Signal Processing. 1997;**144**(3): 145-149. DOI: 10.1049/ip-vis:19971159

[33] Kabir S. A compressed representation of Mid-Crack code with Huffman code. International Journal on Image, Graphics and Signal Processing. 2015;7(10):11-18. DOI: 10.5815/ijigsp.2015.10.02

[34] de Berg M, Cheong O, van Kreveld M, Overmars M. Computational geometry: Algorithms and applications. 3rd ed. Berlin Heidelberg: Springer; 2008. p. 386. DOI: 10.1007/978-3-540-77974-2

[35] Chazelle B, Edelsbrunner H. An optimal algorithm for intersection line segments in the plane. Journal of ACM. 1992;**39**(1):1-54. DOI: 10.1145/147508.147511

[36] Žalik B. Two efficient algorithms for determining intersection points between simple polygons. Computers & Geosciences. 2000;**26**(2):137-151. DOI: 10.1016/S0098-3004(99)00071-0

Chapter 4

# Application of Discrete Mathematics for Programming Discrete Mathematics Calculations

*Carlos Rodriguez Lucatero*

## Abstract

In the discrete mathematics courses, topics, such as the calculation of the element in any position of a sequence of numbers generated by some recurrence relation, calculation of multiplicative inverses in algebraic ring structures modulo a number $n$, obtaining the complete list of combinations without repetition, for which you can take advantage of the computing power of computers and perform such calculations using computer programs in some programming language. The implementations of these calculations can be carried out in many ways and therefore their algorithmic performance can be very varied. In this chapter, I propose to illustrate by means of some Matlab programs, how the use of results of the same discrete mathematics allows to improve the algorithmic performance of said computer programs. Another topic addressed in regular discrete mathematics courses where calculations arise that could become very expensive both in time and in occupied space, if the calculations are implemented directly from the definitions is modular arithmetic. Such calculations can be carried out much more efficiently by making use of results from discrete mathematics and number theory. The application of these ideas will be developed in the following sections of this chapter.

**Keywords:** recurrence relations, algorithms, generating functions, modular arithmetic, Matlab

## 1. Introduction

Discrete mathematics provides very useful calculation tools to enumerate mathematical objects of a certain type, such as the number of graphs that meet a certain particular property or how many regions will be formed within a circle as the number of points that grows, we will join by means of secant lines [1]. The methods and tools of discrete mathematics are of enormous relevance in the field of computer science when one wants to compare different algorithmic solutions to a problem. The goodness of an algorithmic solution must take into account, the use of runtime resources and memory space since they are limited. For this, it is necessary to carry out an analysis of the algorithm and count how many execution steps, such processing takes and how many memory locations it will occupy. When carrying out this analysis, mathematical expressions known as recurrence relationships are obtained, which reflect the behavior at execution time or the memory space occupied by the algorithm. Once the recurrence relationship is obtained, it can be evaluated to estimate, for example, how many executions steps an algorithm will take as the number of input data increases. This generates a succession of values that can eventually be graphed to give us an idea of the temporal complexity of such an algorithm. Such evaluation can be done by hand or it can be implemented, in some programming language available, a function. Normally the direct translation of the mentioned recurrences to a program produces compact and clear recursive routines. The evaluation of such routines is inefficient since many of the recursive calls repeat calculations and it is in this case that it is worth looking for more efficient iterative versions of these calculations. Even more, if possible, we can solve the recurrences to obtain a closed mathematical expression that describes the behavior of the growth in the time of the algorithm. That is where the tools of discrete mathematics acquire special relevance. In algorithm analysis books, one can find many techniques to solve certain types of recurrences and we can illustrate by giving some examples in this chapter. Either way, there are more general discrete mathematics tools for this purpose that we will also try to illustrate with examples. This process of solving recurrences can be seen as a generalization of the solution of classical problems. Some problems consist of finding the next element in a sequence of numbers and for which there are techniques, such as the method of divided differences. Sometimes obtaining the closed solution of a recurrence is not possible, however, discrete mathematics and mathematical analysis allow to define upper and lower bounds as well as approximate calculations of very good quality.

Another topic addressed in discrete mathematics courses, which requires the performance of many calculations, is that of modular arithmetic. Such calculations can be carried out much more efficiently by making use of results from discrete mathematics and number theory. The application of these ideas will be developed in the following sections of this chapter.

## 2. Body of the manuscript

The chapter will have the following structure. In Section 3, we will make a quick revision of some calculations problems in combinatorics. After that, in Section 4, we will talk about where the recurrence relations arise and some methods as well as mathematical tools that are frequently used to solve them. In each example, we will begin by obtaining the mathematical relationship that describes the behavior of the problem to be solved. Once the mathematical relationship is obtained, we will use it to give us an idea of the behavior it describes by evaluating it by means of some function programmed in Matlab. Later, we will obtain the associated mathematical expression or a good approximation function that we will evaluate implementing a Matlab function. This will allow us to compare the performance in the time of the evaluation of an expression and the direct application of a recurrence by implementing both in Matlab.

In Section 5, we will describe characteristics or properties of ring algebraic structures on the set of positive integers modulo some number $n$ and directly from the definitions, we will obtain the tables of the two operations of the ring in question. Later, we will obtain the same tables more efficiently from the application of properties and theoretical results of modular arithmetic to illustrate the advantage of making use of the results of modular arithmetic to make calculations of modular arithmetic algorithmically more efficient both in time and space.

## 3. Combinatorics calculation

One of the most relevant topics in discrete mathematics is combinatorial analysis, where, among other things, it is important to calculate the number of all possible linear arrangements of a given size of objects of a set with repetitions or without repetitions. In some cases, the order of appearance of these objects must be taken into account. When the order of the elements is important and it is allowed to have a repetition of them, we are talking about permutations with repetition. If the repetition of elements is forbidden, then we are talking about permutations without repetition. To clarify ideas, suppose that we have a set $A = \{a, b, c, d\}$ and that we want to calculate all possible linear arrangements of 3-position elements allowing repetitions of elements in each of the positions. The cardinality of the set of objects $A$ is 4 and as each of the three positions can be occupied by any of the elements of $A$ then the total number of possible dispositions will be $4 \times 4 \times 4$ that is to say $4^3 = 64$. The resulting list of possibilities appear in **Table 1**.

This table was generated by the following Matlab function.

```
function y = enumeraPCR(A)
% Autor: Carlos Rodriguez Lucatero
[m,n]=size(A);
cuenta=0;
  for i=1:n
    for j=1:n
        for k=1:n
          fprintf('%s,%s,%s \n',A(1,i),A(1,j),A(1,k));
          cuenta=cuenta+1;
        end
    end
  end
y=cuenta;
end
```

If the repetitions of elements of $A = \{a, b, c, d\}$ are not allowed the linear dispositions of three positions of those elements have four possibilities for the first position, three possibilities for the second position, and two possibilities for the third position, giving as a total number of linear dispositions $4 \times 3 \times 2 = 24$. The listing of those dispositions can be seen in **Table 2**.

| | | | | | | | |
|---|---|---|---|---|---|---|---|
| a,a,a | a,a,b | a,a,c | a,a,d | a,b,a | a,b,b | a,b,c | a,b,d |
| a,c,a | a,c,b | a,c,c | a,c,d | a,d,a | a,d,b | a,d,c | a,d,d |
| b,a,a | b,a,b | b,a,c | b,a,d | b,b,a | b,b,b | b,b,c | b,b,d |
| b,c,a | b,c,b | b,c,c | b,c,d | b,d,a | b,d,b | b,d,c | b,d,d |
| c,a,a | c,a,b | c,a,c | c,a,d | c,b,a | c,b,b | c,b,c | c,b,d |
| c,c,a | c,c,b | c,c,c | c,c,d | c,d,a | c,d,b | c,d,c | c,d,d |
| d,a,a | d,a,b | d,a,c | d,a,d | d,b,a | d,b,b | d,b,c | d,b,d |
| d,c,a | d,c,b | d,c,c | d,c,d | d,d,a | d,d,b | d,d,c | d,d,d |

**Table 1.**
*Complete listing of permutation with repetitions.*

| | | | | | |
|---|---|---|---|---|---|
| a,b,c | a,b,d | a,c,b | a,c,d | a,d,b | a,d,c |
| b,a,c | b,a,d | b,c,a | b,c,d | b,d,a | b,d,c |
| c,a,b | c,a,d | c,b,a | c,b,d | c,d,a | c,d,b |
| d,a,b | d,a,c | d,b,a | d,b,c | d,c,a | d,c,b |

**Table 2.**
*Complete listing of the permutations without repetitions.*

This last table was generated by the following Matlab function.

```
function [y,lista] = ListaPSR(A,k)
% Autor: Carlos Rodriguez Lucatero
[m,n]=size(A);
r=factorial(n)/factorial(n-k);
y(1,1:k)=A(1,1:k);
s=1;
while r-1 > 0
    j=n-1;
    while((A(1,j) >= A(1,j+1)) & (j>0))
        j=j-1;
    end
    l=n;
    while ((A(1,j) >= A(1,l)) & (l>j))
        l=l-1;
    end
    temp=A(1,j);
    A(1,j)=A(1,l);
    A(1,l)=temp;
    t=j+1;
    l=n;
    while (t<l)
        temp=A(1,t);
        A(1,t)=A(1,l);
        A(1,l)=temp;
        t=t+1;
        l=l-1;
    end
    s=s+1;
    y(s,1:k)=A(1,1:k);
    r=r-1;
end
lista=y;
end
```

In general, if the cardinality of the set of objects is $n$ and the number of possible positions within the linear arrangement is $k$ then the total of possible permutations with repetition would be equal to $n^k$. In the event that repetitions of elements are not allowed in the arrangement, the first position can be occupied by any of the $n$ elements of the set of objects. Once the element is assigned to position 1, the next position can be filled by any of the remaining $n-1$ elements and so on until filling the $k$ positions of the linear arrangement. In the general case, the calculation of the total of possible dispositions under the aforementioned conditions would be $n \times (n-1) \times \ldots \times (n-k+1)$. As you can see, this calculation can result in an excessive number of factors and for that reason the $n!$ symbol is introduced, which is equal to $n \times (n-1) \times (n-2) \ldots \times 3 \times 2 \times 1$. Thus, the calculation of the total number of permutations without repetition of $n$ objects taken from $k$ in $k$ is calculated with the following formula

$$\frac{n!}{k!(n-k)!}. \tag{1}$$

The factorial can be defined as a recurrence relation. Every recurrence relationship has a stop condition and a recursive step. To do this, it is necessary to define both parts of the recurrence. For this, we agree that $0! = 1$ which constitutes the stop condition and since $n! = n \times (n-1) \times (n-2) \times \ldots \times 3 \times 2 \times 1 = n \times (n-1)!$. So, we can rewrite $n!$ as recurrence as follows

$$\begin{cases} 0! = 1 & \text{for } n = 0. \\ n! = n \times (n-1)! & \text{for } n > 0. \end{cases} \tag{2}$$

The recurrence 2 to calculate the factorial is one of the first recurrence relationships to appear in discrete mathematics. If we try to implement in Matlab a direct translation of the recurrence 2 this could be the following.

```
function y=fact(n)
if n<0
     disp('argument should be positive');
     y=-1;
elseif n==0
     y=1;
  else
     y=n*fact(n-1);
  end
end
```

Such an implementation is compact and elegant but has the disadvantage that it can easily overwhelm the system's recursive call stack, which is why it is worth looking for an equivalent iterative implementation. Fortunately, it is known that recursive routines whose last instruction is a recursive call can always be translated into an iterative version. Therefore, we propose the following iterative implementation.

```
function y= factI( n )
acum=1;
for i=1:n
     acum=acum*i;
end
y=acum;
end
```

Sometimes when calculations have to be made, it may be enough to use good approximations. The calculation of $n!$ can be done in an approximate and efficient way using the Stirling approximation formula whose proof can be consulted in ref. [2] and whose mathematical expression is the following

$$n! \approx \sqrt{2 \cdot \pi} \cdot n^{n+\frac{1}{2}} e^{-n}. \tag{3}$$

This approximation turns out to be quite good and this allows us to calculate $n!$ very efficiently with a computer, avoiding the problem of excessive recursive calls in the case that $n$ is very large. A possible implementation of this approach using Matlab would be the following.

```
function y = factStirling(n)
% Stirling approximation of the factorial
% Feller Vol. I pag 70
$ Autor: Carlos Rodriguez Lucatero
y=sqrt(2*pi)*n^(n+1/2)*exp(-1*n);
end
```

We can compare the algorithmic complexities of these three different functions to calculate $n!$ and we can notice that both the recursive and iterative versions have a time complexity of $O(n)$ while the Stirling approximation version is of $O(1)$ which is more efficient but at the price of the calculation being approximate. This example of the different ways of calculating $n!$ is the first example of how we can benefit from mathematical results to improve the temporal performance of the programs that we implement for this purpose.

## 4. Recurrences calculation and analysis of algorithms

With regard to algorithm efficiency, it is in this topic that some recurrence relationships and methods of solving said recurrences can be found to determine the time behavior of the algorithms in terms of the number of input data. One of the design techniques that allows you to build efficient algorithms is known as *Divide and Conquer*. A well-known problem in computing is that of ordering in ascending order a set of numerical data in disorder. In regular algorithm analysis courses, various algorithmic solutions for the said problem are studied and some of the most efficient algorithmic solutions are designed using the *Divide and Conquer* approach. One of the most famous Divides and Conquer algorithms is the Merge-Sort. The operation of the Merge-Sort starts by dividing the arrangement into subarrays in half and each subarray divides it again, in the same way, proceeding recursively until it reaches a point where it is not possible to further subdivide subarrays, after which it begins to sort by interleaving the subarrays until merging the ordered subarrays of the first partition. A possible Matlab implementation of this algorithm is shown below.

```
function y = mergesort(A,p,r)
if (p < r)
   q=floor((p+r)/2);
   y1=mergesort(A,p,q);
   y2=mergesort(A,q+1,r);
   y=merge(A,y1,y2,p,q,r);
else
   y=A;
end
end
function y = merge(A,y1,y2,p,q,r)
L=y1(1,p:q);
[m1,n1]=size(L);
L(1,n1+1)=99999999;
R=y2(1,q+1:r);
```

```
[m2,n2]=size(R);
R(1,n2+1)=99999999;
i=1;
j=1;
for k=p:r
   if L(i) <= R(j)
      A(k)=L(i);
      i=i+1;
   else
      A(k)=R(j);
      j=j+1;
   end
end
y=A;
end
```

When analyzing the Merge-Sort, a recurrence of the time or number of steps in total that it takes depending on the size of the input is obtained, which we will denote as $T(n)$. The recurrence obtained as a product of the analysis of this algorithm can have the following form

$$\begin{cases} T(1) = 1 & \text{for } n = 1. \\ T(n) = 2T\left(\frac{n}{2}\right) + n & \text{for } n > 1. \end{cases} \tag{4}$$

To solve the recurrence relation, four several methods, such as substitution, recurrence tree, or iteration of recurrence can be applied [3]. Here we will use the iteration method of recurrence. To simplify the problem, we will assume that the size of the array is $n = 2^m$, that is, it is a power of 2. Taking into account the above, we can operate a variable change in recurrence 4 which is rewritten as follows

$$\begin{cases} T(2^0) = 1 & \text{for } m = 0. \\ T(2^m) = 2T(2^{m-1}) + 2^m & \text{for } m > 0. \end{cases} \tag{5}$$

We iterate over the recurrence 5 we obtain the following expression

$$T(2^m) = 2^m + 2^m + 2^m + \dots + 2^m = m \times 2^m. \tag{6}$$

We know that $n = 2^m$ and therefore that $m = \log_2(n)$. Then we can make the change of variable in expression 5 and obtain the solution to recurrence 4 that would have the following form

$$T(n) = n \log_2(n). \tag{7}$$

Recurrence relationships are present in undergraduate courses in algorithmic analysis, mathematical thinking, or discrete mathematics. One of the topics that are usually addressed in the courses of mathematical thought is that of sequences of numbers and the detection of patterns of behavior of these sequences to illustrate the mathematical process of discovering the properties of sequences of numbers. Typical examples consist of presenting a sequence of integer values and inferring what is the next number in the sequence. In classic mathematical thinking, textbooks such as [4] systematic methods are exposed to solve this type of mathematical puzzles, such as the method of successive differences. Normally these types of exercises remain at the point of discovering the next element in the sequence, but you can go further in the problem and try to discover the mathematical expression that allows obtaining the term of a sequence of numbers in an arbitrary position. It is there where we can resort to the recurrence relations from which a recursive program can be implemented to obtain elements of the numerical sequence in any given position or even more use the tools of discrete mathematics to solve said recurrences to obtain in computationally efficient way elements in arbitrary positions in a sequence of numbers. Next, we will show these ideas with an example with numerical sequences in which we will obtain the following element of the sequence by the method of successive differences, then we will obtain some recurrence relation from which we will implement a function in Matlab that allows us to obtain any element of the sequence given the position and finally, we will solve the recurrence in question to obtain a mathematical formula that represents the form of the n-th term of the sequence from which a function will be implemented to Matlab to evaluate said mathematical expression giving the position to obtain the number that occupies that position within the given sequence of numbers.

## 4.1 Numerical sequences and recurrences

There are sequences of numbers known as figurative numbers since they are related to figures of a certain type that are formed by joining a certain number of points with lines. These points and lines can form, for example, triangles, squares, pentagons, or heptagons, and the associated sequences will be called triangular numbers, square numbers, pentagonal numbers, or heptagonal numbers, respectively. In **Figure 1**, we can see how pentagons can be formed from the number of points given in each case.

The sequence associated with the pentagonal numbers is shown in **Table 3**.

Suppose you want to obtain the next element of the numerical sequence, that is, the element $a_5$. For this, we can apply the method of successive differences that consists of taking the first difference between successive elements of the sequence, then the differences of the first successive differences are taken, and so on.

When the successive differences become constant, the process stops and we perform a backward calculation adding the last differences of each level until we reach the calculation of the last element of the original sequence.

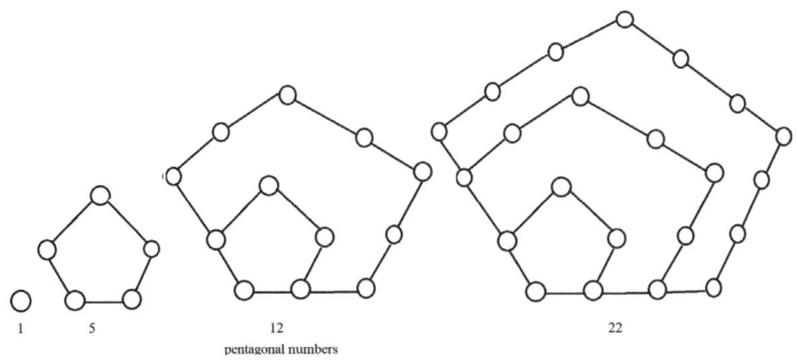


**Figure 1.**
*Pentagonal numbers sequence.*

| $a_0$ | $a_1$ | $a_2$ | $a_3$ | $a_4$ | $a_5$ |
|---|---|---|---|---|---|
| 1 | 5 | 12 | 22 | 35 | |

**Table 3.**
*Pentagonal numeric sequence.*

| $a_0$ | $a_1$ | $a_2$ | $a_3$ | $a_4$ | $a_5$ |
|---|---|---|---|---|---|
| | 5 | 12 | 22 | 35 | **51** |
| 4 | 7 | 10 | 13 | **16** | |
| | 3 | 3 | 3 | **3** | |

**Table 4.**
*Successive differences on the pentagonal sequence.*

The steps performed by this procedure appear in **Table 4**.

The numbers that appear in boldface are generated by the backward calculation mentioned before. This procedure, although correct, can become very cumbersome if instead of wanting to calculate the next element in the sequence, you want to know the value of the element in the sequence in position 30. It is in this case that it would be worthwhile to obtain a mathematical relationship that would allow us to implement a program in any programming language that is available to obtain any element of the numerical sequence in any given position. One way to achieve this goal is to try to discover the recurrence relationship that allows the elements of the sequence to be generated until the element of the sequence in the desired position is reached. This is what we will do next. From the calculations carried out in the successive differences procedure, we can obtain the following first difference relationships between elements of the sequence

$$a_1 - a_0 = 4. \tag{8}$$

$$a_2 - a_1 = 7. \tag{9}$$

$$a_3 - a_2 = 10. \tag{10}$$

$$a_4 - a_3 = 13. \tag{11}$$

$$a_5 - a_4 = 16. \tag{12}$$

From the first differences, we can obtain the following relationships corresponding to the second differences

$$(a_2 - a_1) - (a_1 - a_0) = 3. \tag{13}$$

$$(a_3 - a_2) - (a_2 - a_1) = 3. \tag{14}$$

$$(a_4 - a_3) - (a_3 - a_2) = 3. \tag{15}$$

$$(a_5 - a_4) - (a_4 - a_3) = 3. \tag{16}$$

We can observe the relations related to the second difference; a regular behavior that allows us to establish the following generalization

$$(a_n - a_{n-1}) - (a_{n-1} - a_{n-2}) = 3. \tag{17}$$

From Eq. (17), we can obtain the following difference equation

$$a_n - 2a_{n-1} + a_{n-2} = 3. \tag{18}$$

As can be seen, it is a difference equation, it is non-homogeneous linear second order and with constant coefficients, and for this reason, it requires two initial conditions that are taken from the numerical sequence in question. These initial conditions are $a_0 = 1$ and $a_1 = 5$. Thus, the equation in complete differences would be expressed in the following form

$$a_n - 2a_{n-1} + a_{n-2} = 3, a_0 = 1, a_1 = 5. \tag{19}$$

From Eq. (19), we can obtain the following recurrence

$$\begin{cases} a_0 = 1 & \text{for } n = 0. \\ a_1 = 5 & \text{for } n = 1. \\ a_n = 2a_{n-1} - a_{n-2} + 3 & \text{for } n > 1. \end{cases} \tag{20}$$

Recurrence allows us to directly implement the following recursive function in Matlab.

```
function y = recpentagR(n)
%Author: Carlos Rodriguez Lucatero
if (n==0)
   y=1;
else
   if (n==1)
      y=5;
   else
      y=2*recpentagR(n-1)-recpentagR(n-2)+3;
   end
end
end
```

We can test the routine from Matlab by first calculating an element of the sequence whose value we know, for example $a_5 = 51$. The result of calling the function from the Matlab prompt is the following.

```
>> z=recpentagR(5)
z =
51
```

Now let us try to calculate with this same routine the element of the numerical sequence at position 30, that is, $a_{30}$. The result is shown below.

```
>> z=recpentagR(30)
z =
   1426
```

If we wanted to calculate $a_{50}$ with this recursive routine we would realize that the calculation time increases a lot because many of the calculations are recalculated and also there is a risk of overflowing the system stack due to a large amount of recursive calls. Fortunately, we can reprogram an iterative version of this function and it is the one shown below.

```
function y = recpentagI(n)
%Author: CRL
a0=1;
a1=5;
if (n==0)
    y=a0;
else
    if (n==1)
        y=a1;
    else
        an=0;
        for i=2:n
            an=2*a1-a0+3;
            a0=a1;
            a1=an;
        end
        y=an;
    end
end
end
```

We will test this routine by first executing it for the known value $a_5 = 51$, then for the value $a_{30}$ whose value obtained with the recursive version was 1426 and finally, we will obtain the value $a_{50}$ as well as the $a_{100}$ value.

```
>> z=recpentagI(5)
z =
     51
>> z=recpentagI(30)
z =
     1426
>> z=recpentagI(50)
z =
     3876
>> z=recpentagI(100)
z =
     15251
```

The runtime improvement of the iterative version over the recursive version is truly impressive. However, the execution time of a routine can be further improved to carry out this calculation by resorting to discrete mathematics tools to solve the recurrence and implement a routine that the only thing that does is evaluate in the given position the mathematical formula obtained from the solution of the recurrence. Discrete mathematics provides us with many methods to solve non-homogeneous linear difference equations, such as the one we will solve, associated with recurrence relationships. This type of difference equations can be obtained by methods, such as the iteration of recurrence, the characteristic polynomial method, or the generating function method. In this subsection, we will resolve the recurrence by applying several of these methods for illustrative purposes. For more details on solving recurrences using these methods, we recommend consulting [3, 5–8].

### 4.2 Solving a recurrence by iteration

We start with the application of the iteration method of recurrence. For this purpose, we will use the third row of the recurrence relation (20). This equation would be the following

$$a_n = 2a_{n-1} - a_{n-2} + 3. \tag{21}$$

We apply Eq. (21) to the case of $n - 1$ which would give the following equation

$$a_{n-1} = 2a_{n-2} - a_{n-3} + 3. \tag{22}$$

We substitute 22 in 21 and obtain the following equation

$$a_n = 3a_{n-2} - 2a_{n-3} + 2 \cdot 3 + 3. \tag{23}$$

We apply Eq. (21) to the case of $n - 2$ which would give the following equation

$$a_{n-2} = 2a_{n-3} - a_{n-4} + 3. \tag{24}$$

We substitute 24 in 23 and obtain the following equation

$$a_n = 4a_{n-3} - 3a_{n-4} + 3 \cdot 3 + 2 \cdot 3 + 3. \tag{25}$$

We apply Eq. (21) to the case of $n - 3$ which would give the following equation

$$a_{n-3} = 2a_{n-4} - a_{n-5} + 3. \tag{26}$$

We substitute 26 in 25 and obtain the following equation

$$a_n = 5a_{n-4} - 4a_{n-5} + 4 \cdot 3 + 3 \cdot 3 + 2 \cdot 3 + 3. \tag{27}$$

Continuing with this procedure until reaching the base cases and noting that certain regularities appear, such as the presence of a sum of successive natural numbers, we arrive at the following expression

$$a_n = n \cdot a_1 - (n-1) \cdot a_0 + 3 \cdot \sum_{i=1}^{n-1} i. \tag{28}$$

Substituting $a_0 = 1, a_1 = 5$ and applying Gauss's formula to the summation in Eq. (28), we obtain the following solution

$$a_n = \frac{3}{2} \cdot n^2 + \frac{5}{2} \cdot n + 1. \tag{29}$$

### 4.3 Solving a recurrence by the characteristic polynomial method

We can try another method of solving the recurrence to illustrate another tool provided by discrete mathematics. In this case, we will use the characteristic polynomial method. The difference equation that we are going to solve is linear inhomogeneous with coefficients, which makes it capable of being solved by this method.

These methods are closely related to the methods of solving differential equations of the same type. The difference equation that we are going to solve is the following

$$a_n - 2a_{n-1} + a_{n-2} = 3, a_0 = 1, a_1 = 5. \tag{30}$$

The theory on the solution of equations in non-homogeneous linear differences says that the general solution is composed of a solution of the homogeneous equation plus a particular solution related to the non-homogeneous part. The homogeneous equation associated with Eq. (30) would be the following

$$a_n - 2a_{n-1} + a_{n-2} = 0. \tag{31}$$

A characteristic polynomial is associated with the homogeneous Eq. (31) that is obtained by substituting a possible solution in the difference equation. Said possible solution has the form $a_n = c \cdot r^n$ where $c$ is an arbitrary constant if we substitute said possible solution in 31 we obtain the following polynomial

$$r^2 - 2r + 1 = 0. \tag{32}$$

The roots of the characteristic polynomial 32 are $r_1 = 1$ and $r_2 = 1$, that is, they are real and repeated roots and since the solutions of a difference equation must be linearly independent, the form of the solution of the homogeneous difference equation would be the following

$$a_n = c_1 \cdot 1^n + c_2 \cdot n \cdot 1^n. \tag{33}$$

Now we proceed to solve the particular equation whose solution must be linearly independent of the solutions of the associated homogeneous equation. We note that the right-hand side of the difference Eq. (30) is$f(n) = 3$ which is equivalent to$f(n) = 3 \cdot 1^n$. So the form of the particular solution would be the following

$$a_n = A \cdot n^2. \tag{34}$$

If we evaluate the equation in differences 30 the solution 34 we obtain the following expression

$$A \cdot n^2 - 2 \cdot A \cdot (n-1)^2 + A \cdot (n-2)^2 = 3 \cdot 1^n. \tag{35}$$

After algebraically simplifying Eq. 35 we deduce that $A = \frac{3}{2}$, so the general solution would have the form

$$a_n = c_1 \cdot 1^n + c_2 \cdot n \cdot 1^n + \frac{3}{2} \cdot n^2. \tag{36}$$

Applying on 37 $a_0 = 1$ we deduce that $c_1 = 1$. Applying on 37 $a_1 = 5$ we get $c_2 = \frac{5}{2}$ In this way we obtain that the general solution is

$$a_n = 1 + \frac{5}{2} \cdot n + \frac{3}{2} \cdot n^2. \tag{37}$$

We can verify that solution 37 coincides with solution 29. Finally, we can use this solution to implement a Matlab function to perform this calculation and obtain the value of an element of the sequence given the position. The Matlab function would be the following.

```
function y = recpentag(n)
% Author: Carlos Rodriguez Lucatero
y=1+(5/2)*n+(3/2)*(n^2);
end
```

For being convinced about the correctness of the solution, we can evaluate this Matlab function for the same values used before and we obtain the following results.

```
>> y = recpentag(5)
y =
     51
>> y = recpentag(10)
y =
     176
>> y = recpentag(30)
y =
     1426
>> y = recpentag(50)
y =
     3876
>> y = recpentag(100)
y =
     15251
```

## 4.4 Solving a recurrence by generating function method

Another powerful method of discrete mathematics to solve difference Eq. (19) associated with the recurrence 20 is that of ordinary generating functions. Said method consists, in the simplest case, in converting a numerical sequence into an infinite polynomial of a single variable whose coefficients are precisely the elements of the numerical sequence. The reason for doing this transformation is that the algebraic manipulation of these infinite polynomials is relatively simple. For this reason, we define below the concept of the ordinary generating function of a single variable.

**Definition** 1.1 Given a numerical sequence $a_0, a_1, a_2, \ldots, a_k, \ldots$ the function

$$A(z) = \sum_{k \geq 0} a_k z^k. \tag{38}$$

It is called the ordinary generating function (OGF) of the sequence. The notation $[z^k]A(z)$ will be used to refer to the coefficient $a_k$ of the k-th term of the infinite polynomial.

By means of the generating functions, we can map sequences of numbers that often are integers to power series. The coefficient of the n-th adding will, therefore, be related to the n-th element of the associated numerical sequence. For example, in the generating function $\sum_{i=0}^{\infty} z^i = 1 + z + z^2 + z^3 + z^4 + \ldots$, we observe that it is the geometric series that converges if $|z| < 1$ and can be expressed as $\frac{1}{1-z}$. We can also observe that all the terms of the sum have a coefficient of 1, so this generating function is associated with the numerical sequence $1, 1, 1, 1, \ldots$. On the other hand, since the generating functions are infinite polynomials, it is easy to apply the derivative operation to them. Thus, the derivative of the geometric series would be expressible as follows

$$\frac{d}{dz}\left(\frac{1}{(1-z)}\right) = \frac{1}{(1-z)^2} = \frac{d}{dz}\left(1 + z + z^2 + z^3 + z^4 + \ldots\right) = 1 + 2z + 3z^2 + 4z^3 + \ldots \quad (39)$$

As can be seen from 39, the derivative of the geometric series can be related to de succession of natural numbers.

It is also possible to do certain types of algebraic operations on ordinary generating functions that have an impact on the sequence of numerical values associated with the given generating function. For example, if I multiply the generating function of $\frac{1}{(1-z)^2}$ by $z$, we obtain the following effect in the numerical sequence

$$\frac{z}{(1-z)^2} = 0 + z + 2z^2 + 3z^3 + 4z^4 + \ldots \leftrightarrow 0, 1, 2, 3, 4, 5, \ldots \quad (40)$$

that is to say that we obtain a shift to the right in the sequence of numbers.

If we apply the derivative to Eq. (40), we obtain the following relationship

$$\begin{aligned} \frac{d}{dz}\left(\frac{z}{(1-z)^2}\right) &= \frac{d}{dz}\left(0 + z + 2z^2 + 3z^3 + 4z^4 + \ldots\right) = \frac{z+1}{(1-z)^3} \\ &= 1 + 2^2 z + 3^2 z^2 + 4^2 z^3 + 5^2 z^4 + \ldots \end{aligned} \quad (41)$$

then we have

$$\frac{z+1}{(1-z)^3} = 1 + 2^2 z + 3^2 z^2 + 4^2 z^3 + 5^2 z^4 + \ldots \leftrightarrow 1^2, 2^2, 3^2, 4^2, \ldots \quad (42)$$

We have exemplified the following relationships between numerical sequences and generating functions

$$1, 1, 1, 1, \ldots \leftrightarrow \frac{1}{1-z} \quad (43)$$

$$1, 2, 3, 4, \ldots \leftrightarrow \frac{1}{(1-z)^2}. \quad (44)$$

$$0, 1, 2, 3, 4, \ldots \leftrightarrow \frac{z}{(1-z)^2} \quad (45)$$

$$1^2, 2^2, 3^2, 4^2, \ldots \leftrightarrow \frac{z+1}{(1-z)^3} \quad (46)$$

$$0^2, 1^2, 2^2, 3^2, 4^2, \ldots \leftrightarrow \frac{z(z+1)}{(1-z)^3} \quad (47)$$

After all those examples, we can be convinced that it is possible to establish relationships between sequences of numbers and generating functions, as shown in **Table 5**.

| | |
|---|---|
| $1, 1, 1, 1, 1, \ldots$ | $\frac{1}{1-z} = \sum_{n \geq 0} z^n$ |
| $0, 1, 2, 3, 4, \ldots, n, \ldots$ | $\frac{z}{(1-z)^2} = \sum_{n \geq 1} n z^n$ |
| $0, 0, 1, 3, 6, , 10, \ldots, \binom{n}{2}, \ldots$ | $\frac{z^2}{(1-z)^3} = \sum_{n \geq 2} \binom{n}{2} z^n$ |
| $0, 0, \ldots, 0, 1, m+1, , 10, \ldots, \binom{n}{m}, \ldots$ | $\frac{z^m}{(1-z)^{m+}} = \sum_{n \geq m} \binom{n}{m} z^n$ |
| $1, m, \binom{m}{2}, \ldots, \binom{m}{n}, \ldots, m, 1$ | $(1+z)^m = \sum_{n \geq 0} \binom{m}{n} z^n$ |
| $1, m+1, \binom{m+2}{2}, \ldots, \binom{m+3}{3}, \ldots$ | $\frac{1}{(1-z)^{m+1}} = \sum_{n \geq 0} \binom{n+m}{n} z^n$ |
| $1, 0, 1, 0, \ldots, 1, 0, \ldots$ | $\frac{1}{(1-z)^2} = \sum_{n \geq 0} z^{2n}$ |
| $1, c, c^2, c^3, c^4, \ldots, c^n, \ldots$ | $\frac{1}{1-cz} = \sum_{n \geq} c^n z^n$ |
| $1, 1, \frac{1}{2!}, \frac{1}{3!}, \frac{1}{4!}, \ldots, \frac{1}{n!}, \ldots$ | $e^x = \sum_{n \geq 0} \frac{z^n}{n!}$ |
| $0, 1, \frac{1}{2}, \frac{1}{3}, \frac{1}{4}, \ldots, \frac{1}{n}, \ldots$ | $\ln\left(\frac{1}{1-z}\right) = \sum_{n \geq 1} \frac{z^n}{n}$ |
| $0, 1, 1+\frac{1}{2}, 1+\frac{1}{2}+\frac{1}{3}, \ldots, H_n, \ldots$ | $\frac{1}{1-z} \ln\left(\frac{1}{1-z}\right) = \sum_{n \geq 1} H_n z^n$ (where $H_n$ is the harmonic series) |
| $0, 0, 1, 3\left(\frac{1}{2}+\frac{1}{3}\right), 4\left(\frac{1}{2}+\frac{1}{3}+\frac{1}{4}\right) \ldots$ | $\frac{z}{(1-z)^2} \ln\left(\frac{1}{1-z}\right) = \sum_{n \geq 0} n(H_n - 1) z^n$ |

**Table 5.**
*Table of sequences and ordinary generating functions.*

For more detailed information on recurrences that appear in the analysis of algorithms as well as generating functions, we recommend the excellent book [5].

After this brief summary on ordinary generating functions, we can solve the difference Eq. (19) using generating functions and that is just what we will do next. The difference equation to solve is

$$a_n - 2a_{n-1} + a_{n-2} = 3, a_0 = 1, a_1 = 5. \tag{48}$$

We know from definition 38 that a generating function is expressed mathematically as

$$A(z) = \sum_{k \geq 0} a_k z^k. \tag{49}$$

and we apply this operator to the difference Eq. (48) keeping the same summation index for the terms of the difference Eq. (48) obtaining the following expression

$$\sum_{n \geq 2} a_n z^n - 2\sum_{n \geq 2} a_{n-1} z^n + \sum_{n \geq 2} a_{n-2} z^n = 3\sum_{n \geq 2} z^n. \tag{50}$$

Because of the index shift, the first summation is $A(z) - a_0 - a_1 z$. In the second term of the left hand of Eq. (50) the index of the coefficient and the power of $z$ being different; so, we can factorize $z$ and take into account the shift in the summation index and obtain the term $2z(A(z) - a_0$. The difference between the summation index and the power of $z$ in the third term of the left hand of 50 can be arranged by factoring $z^2$ and in that case, we obtain the term $z^2 A(z)$. The right hand of Eq. (50) is a geometric series but given the shift on the index of the summation, it is necessary to subtract the two first terms of the geometric series giving, as a result, the term $3\left(\frac{1}{1-z} - 1 - z\right)$. Then taking into account these remarks, we obtain the next equation

$$A(z) - 1 - 5z - 2zA(z) - 2z + z^2A(z) = 3\left(\frac{1}{1-z} - 1 - z\right). \tag{51}$$

Simplifying Eq. (51) we get the next equation

$$A(z)\left(1 - 2z + z^2\right) - 1 - 3z = 3\left(\frac{z^2}{1-z}\right). \tag{52}$$

The left hand of 53 can be algebraically simplified obtaining the following expression

$$A(z)(1-z)^2 = 1 + 3z + 3\left(\frac{z^2}{1-z}\right). \tag{53}$$

Simplifying de left hand of Eq. (53) we get the equation

$$A(z) = \frac{1}{(1-z)^2} + 3\frac{z}{(1-z)^2} + 3\left(\frac{z^2}{(1-z)^3}\right). \tag{54}$$

We apply the $[z^n]$ operator to Eq. (54) with the purpose of obtaining the coefficient of the nth addend of the sum that will correspond to the element in the position $n$ of the sequence of numbers studied, arriving at the equation

$$a_n = [z^n]A(z) = [z^n]\left(\frac{1}{(1-z)^2}\right) + 3[z^n]\left(\frac{z}{(1-z)^2}\right) + 3[z^n]\left(\frac{z^2}{(1-z)^3}\right). \tag{55}$$

and then to the final result

$$a_n = (n+1) + 3n + 3\binom{n}{2} = 4n + 1 + \frac{3n^2}{2} - \frac{3n}{2} = 1 + \frac{5n}{2} + \frac{3n^2}{2}. \tag{56}$$

As you can see the expressions (29), (37), and (56) are the same.

## 5. Modular arithmetical calculations

In discrete mathematics courses at the undergraduate level, topics of modern algebra are addressed where the most elementary algebraic structures are defined, such as the group structure and the ring structure. The algebraic structure to which we are going to devote our attention in this section is the ring structure in the context of modular arithmetic.

This algebraic structure is often used, informally, in arithmetic courses to learn to add and multiply. Formally speaking we can say that this structure is composed of a set of objects that could be in the case of arithmetic the set of integers as well as two operations which are addition and multiplication which is denoted as $(\mathbb{Z}, +, \cdot)$. The two operations must be closed, that is to say, that the results they give must belong to the same set from which the operands are taken. Additionally, the operations must respect certain properties that we will define below.

**Definition** 1.2 (Ring) Let be $R$ nonempty set that has two closed binary operations denoted as $+$ y $\cdot$ . Then $(R, +, \cdot)$ it is a ring if $\forall a, b, c \in R$ the following conditions are met:

a) $a + b = b + a$ (commutative law of $+$).

b) $a + (b + c) = (a + b) + c$ (Associative law of $+$).

c) $\exists z \in R$ such that $a + z = z + a = a, \quad \forall a \in R$ (existence of identity element for $+$).

d) For each $a \in R, \ \exists \ b \in R$ such that $a + b = b + a = z$ (existence of inverse under $+$).

e) $a \cdot (b \cdot c) = (a \cdot b) \cdot c$ (Associative law for $\cdot$).

f) $a \cdot (b + c) = a \cdot b + a \cdot c$ y $(b + c) \cdot a = b \cdot a + c \cdot a$ (Distributive laws for $\cdot$ over $+$).

In order to simplify the notation, the operation $a \cdot b$ can be written as $ab$. The associative properties of both operations as well as the distributivity of one operation over the other can be generalized for the case of more than 2 operands.

The $+$ operation of the ring is commutative but the $\cdot$ operation is not commutative in the 1.2 definitions. We can apply the commutativity property to the $\cdot$ operation and we would obtain the commutative ring structure. Formally the definition of a commutative ring is as follows.

**Definition** 1.3 (Commutative ring) Let be $R$ a nonempty set having two closed binary operations denoted as $+$ and $\cdot$ . Then $(R, +, \cdot)$ is a ring if $\forall a, b, c \in R$ meet the following conditions:

a) If $ab = ba, \forall a, b \in R$ then $R$ is a commutative ring.

b) It is said that the ring $R$ has no proper divisors of zero if for any $a, b \in R, ab = z \Rightarrow a = z \vee b = z$.

c) If an element $u \in R$ such that $u \neq z$ and $ua = au = a, \forall a \in R$ we say that $u$ is a unit or identity element of multiplication and that $R$ is a ring with unity.

Another interesting property that can be added to the list of ring properties is the existence of multiplicative inverse. The definition is the following.

**Definition** 1.4 (Ring with multiplicative inverse) Let $R$ be a ring with the unit $u$ if $a, b \in R$ and $ab = ba = u$ then $b$ is called inverse multiplicative of $a$ and $a$ is called a unit of $R$. (An element $b$ is also a unit of $R$).

Taking into account the existence of the unit property, we can add it to the ring commutative ring and obtain a commutative ring with unity.

**Definition** 1.5 (Commutative ring with unity) Let be $R$ a commutative ring with unity. Then:

a) $R$ is said to have an integer domain if $R$ has no proper divisors of zero.

b) $R$ is said to be a field if each non-zero element of $R$ is a unit.

Rings can have subsets that satisfy the properties of a ring as defined in 1.2 in which case we are talking about subring structures. These algebraic structures can have additive cancelation properties as well as multiplicative cancelation properties and therefore will have symmetric or inverse elements of addition as well as multiplication. Once we have understood the theoretical part of the rings we are better positioned to tackle the construction and use of finite rings specials and fields. We will start by presenting some results of the modular arithmetic needed for the de notion of operations in modular rings. These notions will be explained below.

### 5.1 Integers mod *n*

In this section, we will review some notions of modular arithmetic that usually appear in introductory courses of the Theory of Numbers and that allow us to understand important concepts, such as divisibility, primality, the greatest common divisor operation as well as the algorithm of Euclid to calculate it efficiently. Let us start with establishing the meaning of some mathematical symbols that appear in this context. The expression $d|a$ translates into words like d divides a to what is the same as $\exists k \in \mathbb{Z}$ such that $a = kd$ and $d$ is called divisor of $a$. This means that if we divide $a$ by $d$, the remainder will be 0. From here we can make the following properties and definitions.

**Definition** 1.6 Every integer divides 0. That is $\forall x \in \mathbb{Z}, x|0$.

**Proposition** 1.7 If $a > 0$ and $d|a$ then $|d| \leq |a|$.

**Proposition** 1.8 $d|a$ if and only if $-d|a$.

**Definition** 1.9 A number $x$ is said to be composite if it has divisors other than 1 and $x$.

**Example** 1.10 The number 39 is composite since 3|39 is true and $3 \neq 1$ is true as like $3 \neq 39$.

**Example** 1.11 The divisors of 24 are $\{1, 2, 3, 4, 6, 8, 12, 24\}$. The 1 and 24 are known as trivial divisors.

**Definition** 1.12 The set of divisors of a number $a$ that are neither 1 nor $a$ are called factors of $a$.

**Definition** 1.13 A number prime is here the one that has no factors.

**Example** 1.14 The numbers $\{1, 3, 5, 7, 11, 13\}$ are prime.

**Definition** 1.15 A number that is not prime is said to be a composite number.

**Theorem** 1.16 (Division Theorem) For any $a, n \in \mathbb{Z}$ there exist $q, r \in \mathbb{Z}$ única such that $a = qn + r$ where $0 \leq r < n$.

$q = \lfloor \frac{a}{n} \rfloor$ is called the quotient and $r = a \bmod n$ is called the remainder. Then $n|a$ if $a \bmod n$ is equal to 0. From this, in conjunction with theorem 16, it can be stated that $a = \lfloor \frac{a}{n} \rfloor n + a \bmod n$ or equivalently that $a - \lfloor \frac{a}{n} \rfloor n = a \bmod n$.

**Definition** 1.17 If $(a \bmod n) = (b \bmod n)$ then we say that $a \equiv b \bmod n$ if they have the same remainder $a$ and $b$ when divided by $n$ is to say that $a \equiv b \bmod n \Leftrightarrow n|(b - a)$.

When it is true that $(a \bmod n) = (b \bmod n)$ we will denote it as $a \equiv b \bmod n$ and the equivalence classes that this relation generates will be expressed as $[a]_n = \{a + kn | k \in \mathbb{Z}\}$. In general, we can say the congruence relation modulo some number $n$ partitions to the set of integers in equivalence classes by the remainder left when divided by a number $n$ which we express as $[\mathbb{Z}]_n = \{[a]_n : 0 \leq a \leq n - 1\}$.

### 5.2 Common divisors and greatest common divisor

Let us start by defining the concept of a common divisor.

**Definition** 1.18 $d$ is a common divisor of the numbers $a$ and $b$ if $d|a$ and $d|b$ hold.

Common divisors have the following properties.

**Proposition** 1.19 $d|(a + b)$ and $d|(a - b)$.

**Proposition** [1.20] If $d|a$ and $d|b$ then $d|(ax + by)$ $\forall x, y \in \mathbb{Z}$. That is, $d$ divides every linear combination of $a$ and $b$.

**Proposition** 1.21 If $a|b$ then $|a| \leq |b|$ or $b = 0$.

**Proposition** 1.22 If $a|b$ and $b|a$ hold then $a = +b$ or $a = -b$.

**Definition** 1.23 The greatest common divisor of $a$ and $b$ denoted as $gcd(a, b)$. (greatest common divisor) is: $\gcd(a, b) = \max\{d : d|a \ y \ d|b\}$.

**Example** 1.24 $gcd(24, 30) = 6, gcd(5, 7) = 1, gcd(0, 9) = 0$.

**Proposition** 1.25 If $a$ and $b$ are not both equal to 0, then $gcd(a, b)$ is an integer between 1 and $\min(|a|, |b|)$.

**Definition** 1.26 $gcd(0, 0) = 0$.

**Definition** 1.27 $a$ and $b$ are relatively prime if $gcd(a, b) = 1$.

**Proposition** 1.28 $gcd(a, b) = gcd(b, a)$.

**Proposition** 1.29 $gcd(a, b) = gcd(-a, b)$.

**Proposition** 1.30 $gcd(a, b) = gcd(|a|, |b|)$.

**Proposition** 1.31 $gcd(a, 0) = |a|$.

**Proposition** 1.32 $gcd(a, ka) = a, \forall k \in \mathbb{Z}$.

**Theorem** 1.33 If $a$ and $b$ are any integers and are not both equal to 0 then $gcd(a, b)$ is the smallest positive element of the set $\{ax + by : x, y \in \mathbb{Z}\}$ of linear combinations of $a$ and $b$.

**Proof:** Let $s$ be the smallest positive element of the set of such linear combinations of $a$ and $b$, then $s = ax + by$ for some $x, y \in \mathbb{Z}$. Let $q = \lfloor \frac{a}{s} \rfloor$. The equation $a \text{mod} n = a - \lfloor \frac{a}{n} \rfloor$ implies $a \text{mod} s = a - qs = aq(ax + by) = a(1 - qx) + b(-qy)$, so $s \text{mod} s$ is a linear combination of $a$ and $b$. Since $a \text{mod} s < s$ then $a \text{mod} s = 0$ since $s$ is the smallest value that is a linear combination of $a$ and $b$. The above means that $s|a$. Reasoning in a similar way we can prove that $s|b$. Then $s$ is a common divisor of $a$ and $b$, and it also holds that $gcd(a, b) \geq s$ and like $d|a$ and $d|b$ implies that $d|(ax + by)$, so as a consequence of all this $gcd(a, b)|s$ since $gcd(a, b)$ divides any linear combination of $a$ with $b$ and $s$ is a linear combination of $a$ with $b$. But $gcd(a, b)|s$ and $s > 0$ imply that $gcd(a, b) < s$. Combining that is satisfied simultaneously that $gcd(a, b) > s$ and that $gcd(a, b) < s$ we can conclude that $gcd(a, b) = s$.

**Corollary** 1 For any $a, b \in \mathbb{Z}$, if $d|a$ and $d|b$ then $d|gcd(a, b)$.

**Proof:** If $d|a$ and $d|b$ then $\forall x, y \in \mathbb{Z} d|(ax + by)$ and since $gcd(a, b)$ is a linear combination of $a$ and $b$ we can conclude. $\square$.

**Corollary** 2 $\forall a, b, n \in \mathbb{Z}$ and $n \geq 0$ $gcd(an, bn) = n gcd(a, b)$.

**Corollary** 3 $\forall a, b, n \in \mathbb{Z}$ and $a, b, n \geq 0$ if $n|ab$ and $gcd(a, n) = 1$ then $n|b$.

**Theorem** 1.34 $\forall a, b \in \mathbb{Z}^+, gcd(a, b) = gcd(b, a \text{mod} b)$.

**Proof:**

- (demo sketch).
- We first prove that $gcd(a, b)|gcd(b, a \text{mod} b)$.
- Let $d = gcd(a, b)$ then $d|a$ and $d|b$.
- We know that $(a \text{mod} b) = a - qb$ where $q = \lfloor \frac{a}{b} \rfloor$.
- Since $(a \text{mod} b)$ is a linear combination of $a$ and $b$.
- from the above and the fact that $d|a$ and $d|b$ this implies that $d|(ax + by)$ then $d|b$ and $d|(a \text{mod} b)$ where $gcd(a, b)|gcd(b, a \text{mod} b)$.
- As a second part, it is proved in a similar way that $gcd(b, a \text{mod} b)|gcd(a, b)$.
- If both $gcd(a, b)|gcd(b, a \text{mod} b)$ and $gcd(b, a \text{mod} b)|gcd(a, b)$ hold we can conclude that $\forall a, b \in \mathbb{Z}^+, gcd(a, b) = gcd(b, a \text{mod} b)$.

Theorem 1.34 provides us with the mathematical elements to be able to define a recursive algorithm for the calculation of the Greatest Common Divisor is the following.

```
Euclid(a,b)
1) if b=0
2) then return a
3) else Euclid(b, a mod b)
```

The following is an example of how Euclid's algorithm works.

**Example** [1.35] *We will calculate the* $gcd(30, 21)$ *applying the Euclidean algorithm that I have just explained. The execution steps are displayed in* **Table 6**.

## 5.3 Modular rings

**Definition** 1.36 Let $n \in \mathbb{Z}^+, n > 1$. For $a, b \in \mathbb{Z}$, tenths that $a$ is congruent to $b$ modulo $n$ and is written as $a \equiv b(\text{mod}n)$, if $n|(ab)$, or equivalently $a = b + kn$ for some $k \ \ in\mathbb{Z}$.

**Example** 1.37 For example:

a) $17 \equiv 2(\text{mod}5)$.
b) $-7 \equiv -49(\text{mod}6)$.

**Theorem** 1.38 The congruence modulo $n$ is an equivalence relation $\mathbb{Z}$.

**Proof:** (Do it as an exercise).

Since an equivalence relation on a set produces a partition of that set, then for $n \geq 2$ the congruence relation modulo $n$ produces a partition of set $\mathbb{Z}$ in the following $n$ equivalence classes:

$$\begin{aligned}
[0] &= \{\ldots, -2n, -n, 0, n, 2n, \ldots\} = \{0 + nx|x \in \mathbb{Z}\} \\
[1] &= \{\ldots, -2n + 1, -n + 1, 1, n + 1, 2n + 1, \ldots\} = \{1 + nx|x \in \mathbb{Z}\} \\
[2] &= \{\ldots, -2n + 2, -n + 2, 2, n + 2, 2n + 2, \ldots\} = \{2 + nx|x \in \mathbb{Z}\} \\
&\vdots \\
[n-1] &= \{\ldots, -n + 1, -1, n - 1, 2n - 1, 3n - 1\ldots\} = \{(n-1) + nx|x \in \mathbb{Z}\}
\end{aligned}$$

By the division algorithm, we know that $\forall t \in \mathbb{Z}, t = qn + r$ where $0 \leq r < n$, so $t \in [r]$ or also means that $[t] = [r]$. We use $\mathbb{Z}$ to denote $\{[0], [1], [2], \ldots [n-1]\}$ or when there is no ambiguity we also use 'in $\mathbb{Z}_n = \{0, 1, 2, \ldots, n-1\}$.

We can define closed operations addition and pipeline on the set of equivalence classes of $\mathbb{Z}_n$ as $[a] + [b] = [a + b]$ and $[a] \ \ cdot[b] = [a][b] = [ab]$.

**Example** [1.39] If $n = 7$ then $[2] + [6] = [8] = [1]$ and $[2] \cdot [6] = [12] = [5]$.

| | |
|---|---|
| Euclid(30,21) | =Euclid(21,9) |
| | =Euclid(9,3) |
| | =Euclid(3,0) |
| | = 3 |

**Table 6.**
*Execution steps of the Euclid's algorithm.*

Before accepting the definitions of the modular operations $[a] + [b] = [a + b]$ and $[a][b] = [ab]$ we must convince ourselves that they are well defined, that is, if $[a] = [c]$ and $[b] = [d]$ then $[a] + [b] = [c] + [d]$ and $[a][b] = [c][d]$. We will prove that these operations are independent of the choice of elements within a class. So, then $[a] = [c] \Rightarrow a = c + sn$ for some $s \in \mathbb{Z}$ and $[b] = [d] \Rightarrow b = d + tn$ for some $t \in \mathbb{Z}$. Then $a + b = (c + sn) + (d + tn) = (c + d) + (s + t)n$ so that $(a + b) \equiv (c + d) \bmod n$ that is $[a + b] = [c + d]$.

In the same way $ab = (c + sn)(d + tn) = cd + (sd + ct + stn)n$ so that $ab \equiv cd \bmod n$, that is $[ab] = [cd]$. This leads us to establish the following theorem.

**Theorem** 1.40 For $n \in \mathbb{Z}^+, n > 1$ under the closed binary operations just defined $\mathbb{Z}_n$ is a commutative ring with the unit $[1]$.

**Proof:** The proof is left as an exercise for the student. What would have to be done for this is to verify that the ring properties hold for the definition of the addition and multiplication operations on $\mathbb{Z}n$ and the properties of the ring $(\mathbb{Z}, +, \cdot)$.

Before continuing with more theoretical results, let us see the tables of the operations of $+$ and $\cdot$ for $\mathbb{Z}_5$ and $\mathbb{Z}_6$. The $\mathbb{Z}_5$ operation tables are displayed in **Table** 7 and operation tables corresponding to $\mathbb{Z}_6$ ring are shown in **Table 8**.

We can see in the tables of the $+$ and $\cdot$ operations of $\mathbb{Z}_5$ all elements other than 0 have an inverse element, which is why this is a field. In the case of the tables of $\mathbb{Z}_6$, as opposed to those of $\mathbb{Z}_5$, only 1 and 5 are elements with inverse (bf units) and 2, 3, 4 are proper divisors of 0. If we obtained the corresponding tables for $\mathbb{Z}_9$, we could see that $3 \cdot 3 = 3 \cdot 6 = 0$, that is there are also proper divisors of 0 which means that it is not enough for $n$ to be an odd number for the set $\mathbb{Z}_n$ to be a field. The latter leads us to establish the following theorem.

**Theorem** 1.41 $\mathbb{Z}_n$ would be a field if and only if $n$ is a prime number.

**Proof:** Let $n$ be a prime, and suppose $0 < a < n$. Then the $\gcd(a, n) = 1$ and since the $\gcd(a, n)$ is a linear combination of $a$ and $n$ that is to say that $\exists s, t \in \mathbb{Z}_n, as + tn = 1$. Therefore $as \equiv 1(\bmod n)$ i.e. $[a][s] = [1]$, which implies that $a$ is an element with inverse (unit) which makes $\mathbb{Z}_n$ a field. Conversely, if $n$ is not a prime then it can be represented as a product of two numbers, that is $n = n_1 n_2$ where $1 < n_1, n_2 < n$ and if

| + | **0** | **1** | **2** | **3** | **4** |
|---|---|---|---|---|---|
| 0 | 0 | 1 | 2 | 3 | 4 |
| 1 | 1 | 2 | 3 | 4 | 0 |
| 2 | 2 | 3 | 4 | 0 | 1 |
| 3 | 3 | 4 | 0 | 1 | 2 |
| 4 | 4 | 0 | 1 | 2 | 3 |
| · | **0** | **1** | **2** | **3** | **4** |
| 0 | 0 | 0 | 0 | 0 | 0 |
| 1 | 0 | 1 | 2 | 3 | 4 |
| 2 | 0 | 2 | 4 | 1 | 3 |
| 3 | 0 | 3 | 1 | 4 | 2 |
| 4 | 0 | 4 | 3 | 2 | 1 |

**Table 7.**
*Operations tables of the ring* $(\mathbb{Z}_5, +, \cdot)$.

| + | **0** | **1** | **2** | **3** | **4** | **5** |
|---|---|---|---|---|---|---|
| 0 | 0 | 1 | 2 | 3 | 4 | 5 |
| 1 | 1 | 2 | 3 | 4 | 5 | 0 |
| 2 | 2 | 3 | 4 | 4 | 0 | 1 |
| 3 | 3 | 4 | 5 | 0 | 1 | 2 |
| 4 | 4 | 5 | 0 | 1 | 2 | 3 |
| 5 | 5 | 0 | 1 | 2 | 3 | 4 |
| · | **0** | **1** | **2** | **3** | **4** | **5** |
| 0 | 0 | 0 | 0 | 0 | 0 | 0 |
| 1 | 0 | 1 | 2 | 3 | 4 | 5 |
| 2 | 0 | 2 | 4 | 0 | 2 | 4 |
| 3 | 0 | 3 | 0 | 3 | 0 | 3 |
| 4 | 0 | 4 | 2 | 0 | 4 | 2 |
| 5 | 0 | 5 | 4 | 3 | 2 | 1 |

**Table 8.**
*Operations tables of the ring* $(\mathbb{Z}_6, +, \cdot)$.

$[n_1] \neq 0$ so as $[n_1] \neq 0$ but $[n_1][n_2] = 0$ means that $Z_n$ is not an integer domain since it has proper divisors of 0 and therefore cannot be a field.

In $\mathbb{Z}_6$ the $[5]$ has an inverse (it is a unit) and on the other hand, $[3]$ is a proper divisor of 0. It is useful and necessary to be able to detect which elements have an inverse (ie they are units) when $n$ is composite. For this, the following theorem is established.

**Theorem** 1.42 In $\mathbb{Z}_n$, $[a]$ has an inverse (it is a unit) if and only if $\gcd(a, n) = 1$.

**Proof:** If $\gcd(a, n) = 1$ the proof will be the same as the theorem??. In the reverse direction, let $[a] \in \mathbb{Z}_n$ and $[a]^{-1} = s$. So $[a][s] = 1$, so then $as \equiv 1(\text{mod} n)$ and $as = 1 + tn$ for some $t \in \mathbb{Z}$ but $1 = as + n(-t) \Rightarrow \gcd(a, n) = 1$.

In the following example, we illustrate the application of Euclid's algorithms to obtain multiplicative inverses of elements of a modular field-type group.

**Example** 1.43 Find $[25]^{-1}$ in $\mathbb{Z}_{72}$. Since $\gcd(25, 72) = 1$ Euclid's gcd algorithm we get the following:

$$72 = 2(25) + 22, 0 < 22 < 25$$

$$25 = 1(22) + 3, 0 < 3 < 22$$

$$22 = 7(3) + 1, 0 < 1 < 3$$

since 1 is the l'ast non-zero remainder, we have

$$1 = 22 - 7(3) = 22 - 7[25 - 22] = -7(25) + 8(22) = -7(25) + 8[72 - 2(25)] = 8(72) - 23(25)$$

but

$1 = 8(72) - 23(25) \Rightarrow 1 \equiv (-23 + 72)(25)(\text{mod} 72)$, so then $[1] = [49][25]$ then $[49] = [25]^{-1}$ in $\mathbb{Z}_{72}$.

For a more detailed exposition of the previous results in modular arithmetic, greatest common divisor and Euclid's algorithm, it is recommended to consult these bibliographical references [3, 6, 9, 10].

After this brief overview of the properties of the algebraic ring structure as well as how to work with it in the context of modular arithmetic, we are ready to discuss the advantage of applying results from discrete mathematics to perform calculations of modular arithmetic efficiently. As already explained in the previous paragraphs of this subsection, we can, from the tables of the modular ring in which we are working, obtain all the multiplicative inverses of the ring as well as all the proper divisors of zero in the event that the ring does not be a field. We can do this by generating the sum and product tables of the ring and going through the multiplication table detecting a box where the value is equal to 1 in the case of multiplicative inverses, or that the box is equal to 0 in the case of proper divisors of zero. This procedure for obtaining the proper divisors of zero could be carried out with the following Matlab program.

```
function Fz = FacPropZero(n)
% Modulo n [S, P] = TabOpMod (n)
% Search on the table P all those factors whose product is equal to 0
% CRL 29/sep/2021
[S,P] = TabOpMod(n);
[r,c] = size(P);
k=1;
for i=1:r
    for j=1:r
        if (P(i,j)==0)
          Fz{1,k}=[i-1,j-1];
          k=k+1;
        end
     end
end
end
```

The procedure for obtaining the ring elements that have multiplicative inverse could be carried out with the following Matlab program.

```
function U = unitsZn(n)
% Modulo n [S,P] = TabOpMod(n)
% Search on the table P all those factors whose product is equal to 1
% CRL 22/sep/2021
[S,P] = TabOpMod(n);
[r,c] = size(P);
k=1;
for i=1:r
      for j=1:r
           if (P(i,j)==1)
               U{1,k}=[i-1,j-1];
               k=k+1;
           end
      end
end
end
```

The results of the execution of the Matlab function for obtaining the list of elements of the modular ring that have multiplicative inverse for the case of $n = 5$ and $n = 6$ are the following:

```
>> U=unitsZn(5)
U =
  1×4 cell array
    {1×2 double}  {1×2 double}  {1×2 double}  {1×2 double}
>> U{1,1}
ans =
      1    1
>> U{1,2}
ans =
    2    3
>> U{1,3}
ans =
    3    2
>> U{1,4}
ans =
    4    4
>> U=unitsZn(6)
U =
    1×2 cell array
       {1×2 double}  {1×2 double}
>> U=unitsZn(6)
U =
    1×2 cell array
       {1×2 double}  {1×2 double}
>> U{1,1}
ans =
    1    1
>> U{1,2}
ans =
    5    5
```

The above Matlab routines look good and bad from an algorithmic perspective. The good side is that since it is a direct translation of the operations table generation of a modular ring, the programming of the routines is relatively simple. The negative aspect is in the fact that the size of the arrays to generate the tables can become very large as the number $n$ with respect to which the module is taken grows a lot, that is, the memory occupied will be of the order of $O(n^2)$. It is at this point that the properties and results of discrete mathematics in the subject of modular arithmetic come to the rescue and allow us to obtain more efficient algorithms for obtaining proper divisors of zero as well as the list of elements of the modular ring that have an inverse multiplicative. The efficient algorithm for obtaining the list of elements of a modular ring with multiplicative inverse and the respective Matlab implementation will make use of discrete mathematics results, such as Euclid's algorithm for obtaining the greatest common divisor, as well as the version extension of this, and that will be called by a routine that solves linear modular equations of the type $ax = b \bmod n$. The multiplicative inverse $x$ of a number $a$ modulo $n$ will correspond to the solution of the modular linear equation $ax = 1 \bmod n$. Based on theorem 34, we can implement the Euclid's algorithm in Matlab as follows.

```
function y = mcdEuclides(a,b)
% Recursive function for the GCD using Euclid's algorithm
% The first parameter musy be bigger than the second parameter
if a < b
   temp=a;
   a=b;
   b=temp;
end
if b == 0
   y=a;
else
      y=mcdEuclides(b, mod(a,b));
end
end
```

Euclid's algorithm can be extended to obtain the coefficients $a$ and $b$ of $x$ and $y$ in the relation $d = gcd(a,b) = ax + by$. This will be used for obtaining multiplicative inverses of a modular ring. The Matlab implementation of the extended Euclid algorithm is as follows.

```
Function [d,x,y] = mcdEuclidesExtend(a,b)
% Extended GCD Euclid's algorithm
if a < b
   temp=a;
   a=b;
   b=temp;
end
if b == 0
   d=a;
   x=1;
   y=0;
else
   [d1,x1,y1]=mcdEuclidesExtend(b, mod(a,b));
   d=d1;
   x=y1;
   y=x1-floor(a/b)*y1;
end
end
```

As already mentioned, obtaining the elements of a modular ring that have a multiplicative inverse can be reduced to the problem of calculating the elements $x$ that satisfy the solution to the modular equation $ax = 1 \bmod n$. The Matlab implementation of the linear modular equation solver will be the following.

```
Function S = ModLinEqSolv(a,b,n)
% Modular linear equation solver
% That have the form ax=b mod n
% Author: Carlos Rodriguez Lucatero 29/sep/2021
[d,x,y] = mcdEuclidesExtend(a,n);
if (mod(b,d)==0)
     x0=mod((y*(b/d)),n);
     for i=0:d-1
          S(i+1)=mod(x0+(i*(n/d)),n);
     end
```

```
else
    S=-1;
end
end
```

If we call this routine with the parameter $b = 1$, it would give us the result of the value of $x$, which is the inverse of $a$ in the relation $ax = b\text{mod}n$, if it exists, or it would return $-1$ to indicate that the element $a$ of the modular ring does not have a multiplicative inverse. The following routine calls the linear modular equation solver routine to get the list of elements of the modular ring that have an inverse.

```
Function L = unitsZnV3(n)
% Get the list of units of a modular ring
% mod n calling the routine that solves linear modular equations
% Autor: Carlos Rodriguez Lucatero 14/Ene/2022
k=1;
for i=1:n
    S = ModLinEqSolv(i,1,n);
    if (S ~=-1)
        L(1,k)=S;
    end
    k=k+1;
end
if (length(L)==0)
    L=-1;
end
end
```

The elements of the list in 0 correspond to the inverse of the element that does not have an inverse and therefore we would be talking about dividing elements proper to zero of the modular ring in question. The elements in the list that are not null are the inverse of the element of the modular ring represented by the position in the list. For instance, in $(\mathbb{Z}_5, +, \cdot)$, we have that $2 \cdot 3 = 1$ and $4 \cdot 4 = 1$.

```
>> L = unitsZnV3(5)
L =
    1   3   2   4
>> L = unitsZnV3(6)
L =
    1  0  0  0  5
```

## 6. Conclusions

In this chapter, we address some calculation problems that take place in discrete mathematics and how we can use the theoretical results provided by discrete mathematics itself to be able to carry out these calculations more efficiently. In this chapter, we were able to verify the great utility of the use of generating functions to carry out some calculations more efficiently. Some interesting applications of the use of generating functions to count graphs with some property can be found in refs. [11, 12]. These counting techniques can later be applied to the famous probabilistic method [13–15]. Generating functions are also very useful for counting the number of possible partitions of integers. An excellent text that addresses this interesting topic is [16].

It is true that computers have not stopped increasing their storage capacity as well as the speed of their processing units. However, these resources are not infinite and there are even calculation-intensive problems in topics, such as combinatorics or modular arithmetic that could quickly exhaust these calculation resources. That is why it is worth taking advantage, when possible, of the theoretical results that discrete mathematics itself provides, to efficiently carry out the calculations that it requires. For this, we take two specific topics of discrete mathematics where that is possible. One is the obtainment of elements in arbitrary positions of a numerical sequence and the other topic was the obtainment of elements with multiplicative inverse of the elements that are proper divisors of zero in a modular ring. We illustrate this by programming functions in Matlab. I hope that the chapter will convince you of the goodness of taking advantage of the mathematical results offered by discrete mathematics to perform efficient calculations in the area of discrete mathematics.

## Acknowledgements

I would like to thank the Universidad Autonoma Metropolitana, Unidad Cuajimalpa for the support they gave me to make and publish this book chapter.

## Author details

Carlos Rodriguez Lucatero
Universidad Autónoma Metropolitana Unidad Cuajimalpa, CDMX, Mexico

*Address all correspondence to: crodriguez@cua.uam.mx

## References

[1] Rodríguez-Lucatero C. The Moser's formula for the division of the circle by chords problem revisited. 2017. Available from: https://arxiv.org/abs/1701.08155v1

[2] Feller W. An Introduction to Probability Theory and its Applications. Vol. I. New York, USA: Wiley and Sons Inc; 1968. pp. 52-53

[3] Cormen TH, Leiserson CE, Rivest RL, Stein C. Introduction to Algorithms. 3d ed. Massachusetts, USA: The MIT Press; 2009

[4] Miller CD, Heeren VE, Hornsby J. Matematica, razonamiento y aplicaciones. USA: Pearson; 2013

[5] Sedgewick R, Flajolet P. An Introduction to the Analysis of Algorithms, Second Printing. USA: Addison-Wesley; 2001

[6] Grimaldi RP. Discrete and Combinatorial Mathematics: An Applied Introduction. 3rd ed. USA: Addison-Wesley; 1994

[7] Graham RL, Knuth DE, Patashnik O. Concretel Mathematics, 6th Printing. USA: Addison-Wesley; 1990

[8] Wilf HS. Generatingfunctionology. 3th ed. Massachusetts, USA: A. K. Peters Ltd.; 2006

[9] Hardy GH, Wright EM. Introduction to the Theory of Numbers. 5th ed. Oxford, UK: Oxford Science Publications, reprinted; 1998

[10] Vinográdov I. Fundamentos de la Teora de números. URSS: Moscu: Editorial MIR; 1977

[11] Harary F, Palmer EM. Graphical Enumeration. New York, NY, USA; London, UK: Academic Press; 1973

[12] Rodríguez-Lucatero C. Combinatorial Enumeration of Graphs. Rijeka: IntechOpen; 2019

[13] Erdös P. Graph theory and probability. Canadian Journal of Mathematics. 1959;**11**:34-38

[14] Alon N, Spencer JH. The Probabilistic Method. 2nd ed. New York Wiley-Interscience; 2000

[15] Rodríguez-Lucatero C, Alarcón L. Use of enumerative combinatorics for proving the applicability of an asymptotic stability result on discrete-time SIS epidemics in complex networks. MDPI Mathematics Open access Journal. 2019;7(1). DOI: 10.3390/math7010030

[16] Andrews GE. In: Rota GC, editor. The Theory of Partitions Encyclopedia of Mathematics and its Applications. Vol. 2. USA: Addison-Wesley; 1976

Chapter 5

# A Criticality Study of Fast Critical Experimental Benchmarks Using MCNP Code to Qualifying Different Evaluations

*Sanae El Ouahdani, Hamid Boukhal, El Mahjoub Chakir, Ahmed Gaga, Houda Elyaakoubi, Mustapha Makhloul, Abdelaziz Ahmed, Abdessamad Didi and Mohamed Bencheikh*

## Abstract

In this chapter we present our MCNP modeling, concerning fast critical experimental benchmarks, about qualifying our libraries of cross-sections deduced from the evaluations ENDF/B-VII, JEFF-3.1, JENDL-3.3, JENDL-4 processed by the code NJOY. The benchmarks analyzed are characterized by simple geometries which help to have a precise calculation. In our neutron calculation, we used the MCNP code (version 5), the reference code for the neutron transport calculation with the Monte Carlo method. It is also very efficient for criticality calculation. The cross-section data for all the isotopes that make up the material of the studied benchmarks are processed in ACE format at 300 K temperature using the NJOY 99.9 modular system. A detailed comparison of the criticality results of our simulation was carried out to highlight the influence of these evaluations on the $k_{eff}$ calculations.

**Keywords:** benchmark, MCNP code, NJOY code, ENDF/B-VII, multiplication factor, criticality

## 1. Introduction

The effective multiplication factor $k_{eff}$ is an important parameter in the design, control and safety of reactors. For safety considerations, the $k_{eff}$ is desired to be very close to one throughout the core life. The calculation of the $k_{eff}$ is rather a complicated problem due to the contributions of different physical phenomena related to the neutron's population change. That is why it is important to validate any reactor calculation tool and any nuclear data library with an accurate prediction of this parameter.

The main objective of the present work is to perform the qualification and analysis of the most recent nuclear data libraries available to the scientific community, in particular; ENDF/B-VII [1], JENDL-4.0 [2], JENDL-3.3 [3], and JEFF-3.1 [4], to check the accuracy of cross-section libraries for the criticality calculations. For this objective a set of critical fast benchmarks highly enriched uranium and with $^{233}U$ and with $^{239}Pu$ fuel rods were used to consider as closely as possible all types of geometries to simulate the criticality coefficient of interest. The continued energy cross sections necessary for the present work were processed by the NJOY system (version 99.9, update 364) [5] in the ACE format. The analysis and the interpretation of the results were reinforced by a comparison study of the parameter with the experimental values excerpt from the literature [6]. These experiments have already been analyzed by the Monte Carlo code MCNP using the American nuclear data ENDF/B-V continuous [6] and other codes.

The first part of the paper is reserved to explain the methodology and the materials used. The materials used are the MCNP code and the NJOY code and the JANIS code. The second part cites the characteristics of the different benchmarks selected for the present study. In the third part, we develop our results obtained about the simulation of the $k_{eff}$ parameter and their interpretations concerning the qualification of the used libraries. We finished with a conclusion.

## 2. Methods and materials

### 2.1 MCNP code

The MCNP code [7] (Monte-Carlo N Particle transport), is a code that deals with the transport of neutrons, photons and electrons or coupled/photon/electron by the Monte-Carlo method, including the possibility of calculating the values clean for critical systems. The code deals with an arbitrary three-dimensional configuration with materials in geometric cells delimited by surfaces.

The code takes into account the processed cross-sections, for neutrons, all reactions, which are proposed in particular data evaluations (for example, ENDF/B-VI), thermal neutron scattering which is treated both by the model of the free gas and S ($\alpha$, $\beta$), for photons the code takes into account incoherent and coherent diffusions, the possibility of fluorescence emission after the photoelectric effect, absorption in the production of pairs with a local broadcast of annihilation radiation. It can also treat the braking radiation emitted. In this way, MCNP is qualified as a three-dimensional, continuous energy code, thus it has been proven to simulate physical phenomena correctly. The series of important features that make MCNP very flexible and easy to use code is that it includes a powerful general source, criticality source, surface source, geometry and output pointing plotters, a rich collection of variance reduction techniques, the desired result structure called "tally" and has a large collection of cross-section data.

### 2.2 NJOY code

The NJOY nuclear data processing code [5] is a system developed at the Los Alamos laboratory in the USA since 1974. It is a modular code allowing, from the assessments of so-called basic nuclear data, to create specific or multigroup parameters (multi-group cross sections, fission spectra, etc…) because the information contained in these files is, such as it cannot be, exploited directly by the various transport codes MCNP, WIMS, APPOLO, EPRI-CELL… etc. The role of the NJOY system is to process this information and make it usable by these codes. The data processed by this system are then stored in files in a standardized ENDF (Evaluated Nuclear Data File) format.

### 2.3 JANIS

The enormous amount of data stored in the standard ENDF format files as well as the different versions or evaluations do not always allow easy access to the information desired by the user for a particular application. JANIS (Java-based Nuclear Information Software) [8] is a program designed to facilitate the visualization and manipulation of nuclear data. It was developed by the "OECD Nuclear Energy Agency", the "CSNSM-Orsay" and the University of Birmingham as an extension of the JEF-PC program. The main objective of this program is to allow the user to access the numerical values and the graphic representation of the various data without any prior information on the ENDF format. It gives maximum flexibility for the comparison of different types of nuclear data.

## 3. The fast critical benchmarks

### 3.1 The benchmarks

The benchmarks are fixed points of reference used to test the results of modeling and theoretical calculations and to validate nuclear data.

There are two types of benchmarks:

- Theoretical benchmarks: Are exact references with great precision, obtained by solving mathematical equations describing physical phenomena and processes. They are used primarily to conduct a rough and inherent validation of algorithms widely used in computer codes.

- Experimental benchmarks: Are experiments using measuring instruments dedicated to a good description of physical aspects and phenomena. They are mainly used for the qualification of nuclear data.

### 3.2 Characteristics of the fast benchmarks used

The benchmarks analyzed cover different and simple geometries (spherical, cylindrical, and parallelepiped), with or without reflector, and concern the three main fissile nuclei $^{235}U$, $^{233}U$, $^{239}Pu$ in metallic form.

Fast benchmarks use a fast neutron spectrum that covers the energy range greater than 100 keV, and are therefore characterized by very high fission and capture percentages in the fast energy domain.

By way of example, **Tables 1–3** give the average percentages of the flux as well as the fission and capture rates in the following three energy intervals [6]:

- Thermal energy interval, characterized by neutron energies below 0.625 eV.

- Epithermal energy interval between 0.625 eV and 100 keV.

- Fast energy interval, for neutrons with energy greater than 100 keV.

| Benchmarks | <0.625 eV | 0.625 eV–100 keV | >100 keV |
|---|---|---|---|
| HEU-MET-FAST | 0.005% | ≈6.74% | ≈93.255% |
| PU-MET-FAST | 0.00% | ≈4.41% | ≈95.5% |
| U233-MET-FAST | 0% | ≈3.51% | ≈96.48% |

**Table 1.**
*Average percentages of flux in the three energy intervals.*

| Benchmarks | <0.625 eV | 0.625 eV–100 keV | >100 keV |
|---|---|---|---|
| HEU-MET-FAST | 0.75% | 10.035% | 89.215% |
| PU-MET-FAST | 1.64% | ≈5% | 93.36% |
| U233-MET-FAST | 0% | 4.82% | 95.19% |

**Table 2.**
*Average percentages of fissions caused by neutrons in the three energy fields.*

| Benchmarks | <0.625 eV | 0.625 eV–100 keV | >100 keV |
|---|---|---|---|
| HEU-MET-FAST | 0.815% | 23.32% | 75.865% |
| PU-MET-FAST | 5.72% | 25% | 69.31% |
| U233-MET-FAST | 0.00% | 10% | 90% |

**Table 3.**
*Average percentages of neutron capture in the three energy domains.*

## 3.3 Description of the fast benchmarks studied

As we mentioned before, to qualify our cross-section libraries as well as the modeling method, we have chosen a series of critical fast experimental benchmarks which cover different geometries and relate to the three main fissile nuclei $^{235}$U, $^{239}$Pu and $^{233}$U. These benchmarks are derived from the International Handbook of Critical Benchmarks published by the nuclear energy agency AEN [6].

### *3.3.1 Fast benchmarks highly enriched in U-235 (HEU-MET-FAST)*

We have processed a series of 20 highly enriched benchmarks known with HEU-MET-FAST that is chosen carefully with simple geometries. It includes GODIVA, TOPSY, FLATTOP and HEU-MET-FAST-xxx.

**HEU-MF-001**: GODIVA (1950–1959, LANL, USA), sphere containing metallic uranium highly enriched in the isotope $^{235}$U (93.71% wt*).

*wt = mass fraction.

**HEU-MF-002**: TOPSY-8 (1950, LANL, USA), assemblages of different geometry (depending on the case) containing uranium highly enriched in the $^{235}$U isotope (93.55% wt) reflected by natural uranium, (6 cases).

**HEU-MF-003**: ORALLOY (1950, LANL, USA), spherical assemblies containing metallic uranium highly enriched in $^{235}$U (93.5% wt), reflected by reflectors of different types and thicknesses depending on the case (12 cases): seven spheres are reflected by 5.08, 7.62, 10.16, 12.7, 17.78, 20.32, 27.94 cm of natural uranium, four spheres are reflected by 4.826, 7.366, 11.43, 16.51 cm of Tungsten carbon, one sphere is reflected by 20.32 cm of nickel.

**HEU-MF-028**: FLATTOP-25 (1964–1966, LANL, USA), a sphere containing metallic uranium highly enriched in the $^{235}$U isotope (93.24% wt) reflected by natural uranium.

### *3.3.2 Fast benchmarks in U-233 (U233-MET-FAST)*

**U233-MF-001:** JEZEBEL-23 (1961, LANL, USA), sphere containing metallic uranium highly enriched in the $^{233}$U isotope (98.11% wt).

**U233-MF-002:** (1958, LANL, USA), sphere containing metallic uranium highly enriched in the $^{233}$U isotope (50.59% wt) reflected by a layer of 235 U, (2 cases, in both cases the mass varies critical).

**U233-MF-003:** (1958, LANL, USA), sphere containing metallic uranium highly enriched in the $^{233}$U isotope (98.89% wt) reflected by natural uranium, (2 cases, in both cases the critical mass varies).

**U233-MF-004:** (1958, LANL, USA), sphere containing metallic uranium highly enriched in the $^{233}$U isotope (98.2% wt) reflected by tungsten, (2 cases, in the 2 cases varies the critical fatigue and the reflector thickness).

**U233-MF-005:** (1958, LANL, USA), sphere containing metallic uranium highly enriched in the $^{233}$U isotope (98.2% wt) reflected by beryllium, (2 cases, varies the critical mass).

**U233-MF-006:** FLATTOP-23 (1964, LANL, USA), sphere containing metallic uranium highly enriched in the $^{233}$U isotope (98.13% wt) reflected by natural uranium.

### *3.3.3 Fast benchmarks in Pu-239 (Pu-MET-FAST)*

**Pu-MF-001:** JEZEBEL-39 (1950, LANL, USA), metallic plutonium sphere enriched in the $^{239}$Pu isotope (95.17%), (4.5 at% $^{240}$Pu, 1.02 wt% Ga), without a reflector.

**Pu-MF-002:** JEZEBEL-40 (1964, LANL, USA), metallic plutonium sphere enriched in the $^{239}$Pu isotope (20.1 at% $^{240}$Pu, 1.01 wt% Ga), without a reflector.

**Pu-MF-005:** (1958, LANL, USA), metallic plutonium sphere enriched in the $^{239}$Pu isotope (94.76%), reflected by tungsten.

**Pu-MF-006:** FLATTOP-39 (1964–1966, LANL, USA), metallic plutonium sphere highly enriched in the $^{239}$Pu isotope (94.84% wt) reflected by natural uranium.

**Pu-MF-008:** THOR (1960–1961, LANL, USA), metallic plutonium sphere highly enriched in the $^{239}$Pu isotope (94.54% wt), reflected by thorium.

**Pu-MF-009:** (1960, LANL, USA) plutonium metallic sphere highly enriched in the $^{239}$Pu isotope (94.8% wt), reflected by aluminum.

**Pu-MF-010:** DELTA-PHASE (1958, LANL, USA): metallic plutonium sphere highly enriched in the $^{239}$Pu isotope (94.76% wt), reflected by natural uranium.

**Pu-MF-011:** ALPHA-PHASE (1968, LANL, USA), metallic plutonium sphere highly enriched in $^{239}$Pu (94.4% wt) reflected by light water.

**Pu-MF-018:** DELTA-PHASE (1958, LANL, USA), metallic plutonium sphere highly enriched in the $^{239}$Pu isotope (94.7% wt) reflected by beryllium.

**Pu-MF-023:** (1962, VNIIEF, Russia), metallic plutonium sphere highly enriched in the $^{239}$Pu isotope (98.19%), reflected by the graphite.

**Pu-MF-024:** (1964, VNIIEF, Russia), metallic plutonium sphere highly enriched in the $^{239}$Pu isotope (98.19%), reflected by polyethylene.

**Pu-MF-025:** (1964, VNIIEF, Russia), metallic plutonium sphere highly enriched in the $^{239}$Pu isotope (98.19%), reflected by stainless steel (1.55 cm).

**Pu-MF-026:** (1962, VNIIEF, Russia), metallic plutonium sphere highly enriched in the $^{239}$Pu isotope (98.19%), reflected by stainless steel (11.9 cm).

**Pu-MF-027:** (1965, VNIIEF, Russia), metallic plutonium sphere highly enriched in the $^{239}$Pu isotope (89.66%), reflected by polyethylene.

**Pu-MF-028:** (1965, VNIIEF, Russia), spherical assembly in metallic plutonium highly enriched in the $^{239}$Pu isotope (89% wt) reflected by stainless steel.

**Pu-MF-029:** (1965, VNIIEF, Russia), spherical assembly in metallic plutonium highly enriched in the $^{239}$Pu isotope (88% wt), without a reflector.

**Pu-MF-030:** (1965, VNIIEF, Russia), spherical assembly in metallic plutonium highly enriched in the $^{239}$Pu isotope (88% wt) reflected by the graphite.

**Pu-MF-031:** (1965, VNIIEF, Russia), spherical assembly in metallic plutonium highly enriched in the $^{239}$Pu isotope (88% wt) reflected by polyethylene.

**Pu-MF-032:** (1965, VNIIEF, Russia), spherical assembly in metallic plutonium highly enriched in the $^{239}$Pu isotope (88% wt) reflected by stainless steel.

## 4. Results and interpretations

For the calculation of the $k_{eff}$ parameter, we used the MCNP code based on the Monte Carlo method. The Monte Carlo method solves the transport equation in integral form. The latter is based on the random selection of several variables and after the estimation of their mathematical expectation which is equivalent to the value of the physical quantity sought. It simulates the history of each neutron through the different interactions it can have in the media where it propagates.

In the present calcul, we simulated 1500 cycles of 30,000 neutrons each, the first 50 cycles are used to ensure the homogeneity of the source distribution. With this number of simulated stories, all $k_{eff}$ results are obtained with a standard deviation between +/− 9 and +/− 12 pcm.

## 5. Case of fast benchmarks highly enriched in $^{235}$U

The experimental values obtained for the various Benchmarks concerning the effective multiplication factor $k_{eff}$, as well as the average deviations from experience are shown and compared to the experience in **Figures 1** and **2**.

**Figure 1** represents the variation of $k_{eff}$ according to the cases for the fast benchmarks very highly enriched in $^{235}$U, from this figure we notice that for the majority of the cases studied, the ENDF/B-VII and JEFF-3.1 evaluations give results that are in good agreement with experience. The average deviation from experience is in the order of 0.42% for ENDF/B-VII and 0.39% for the JEFF-3.1 evaluation: all the libraries keep the same difference between themselves and the same behavior for the benchmarks reflected by natural uranium, except for the benchmarks from HEU-MF-008 to HEU-MF-011 which are reflected by tungsten carbide and HEU-MF-012 reflected by nickel. We also note that the JENDL-3.3 evaluation underestimates the criticality in most cases, with an average deviation from experience equal to 0.6%. However, there is a marked improvement when upgrading the evaluation from JENDL-3.3 to JENDL-4. Although, we still have an underestimation compared to the other evaluations of JENDL-3.3 and JENDL-4. We notice an overestimation of $k_{eff}$ for all the evaluations

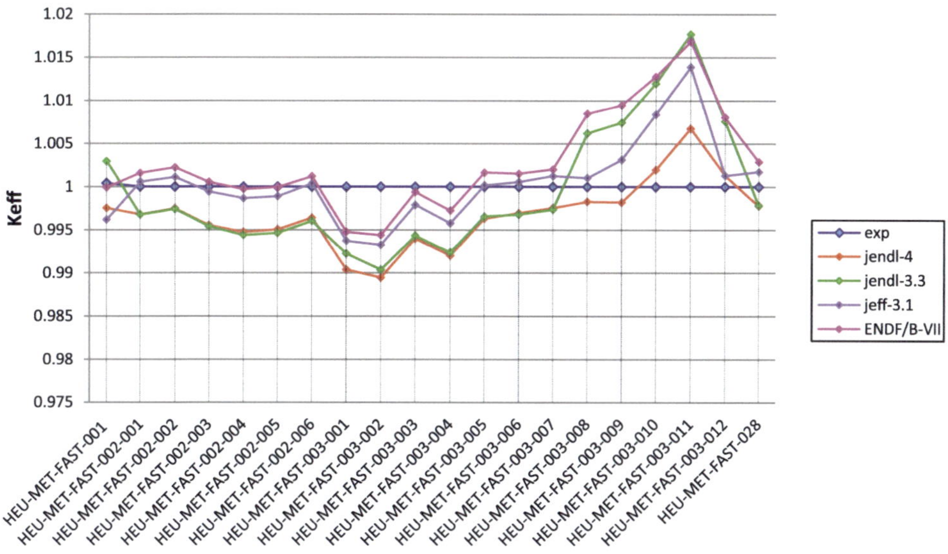


**Figure 1.**
*$k_{eff}$ depending on the case for the fast benchmarks very highly enriched in $^{235}U$.*

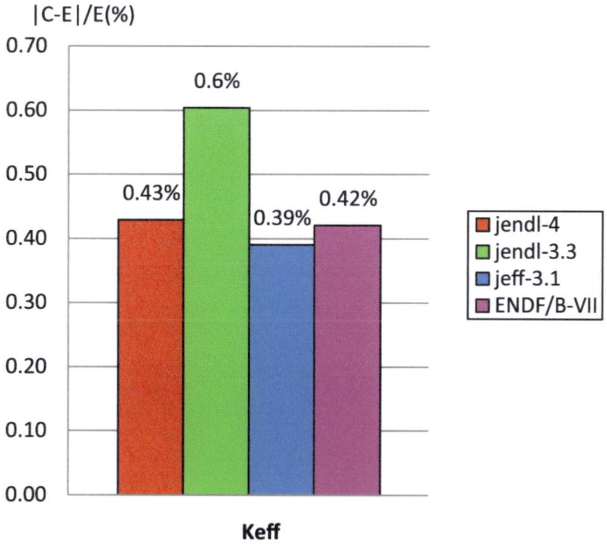


**Figure 2.**
*The |C-E|/E ratio of $k_{eff}$ for each evaluation.*

concerning the benchmarks reflected by the Tungsten carbide which contains the carbon the problem probably stems from a poor underestimation of the carbon capture cross-sections especially in the energy interval of 5 keV to 5 MeV of the capture cross-section where JENDL-4 over estimates ENDF/B-VII and JEFF-3.1.

## 5.1 Fast benchmarks in U-233

**Figures 3** and **4** represent the variation of $k_{eff}$ according to the cases for the fast benchmarks in $^{233}U$ isotope as well as the average deviations from the $k_{eff}$ experiment.

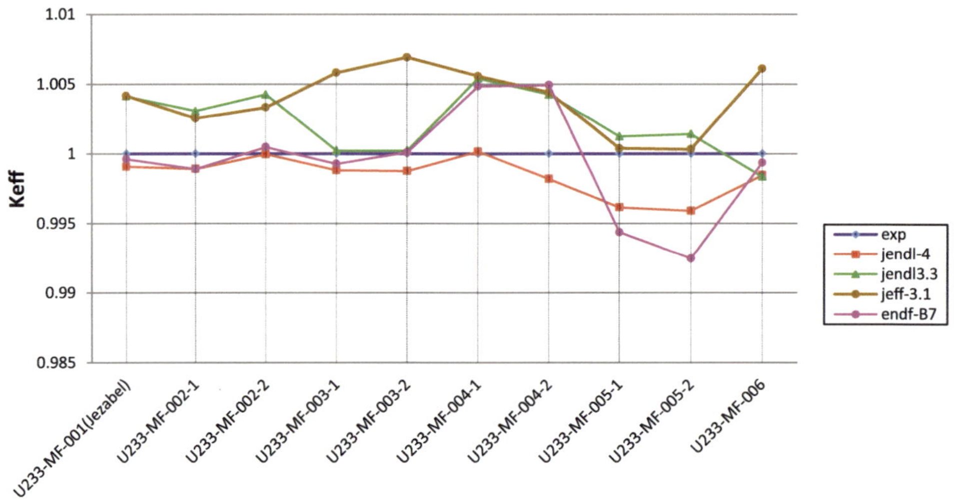


**Figure 3.**
*$k_{eff}$ depending on the case for the fast benchmarks in $^{233}U$.*

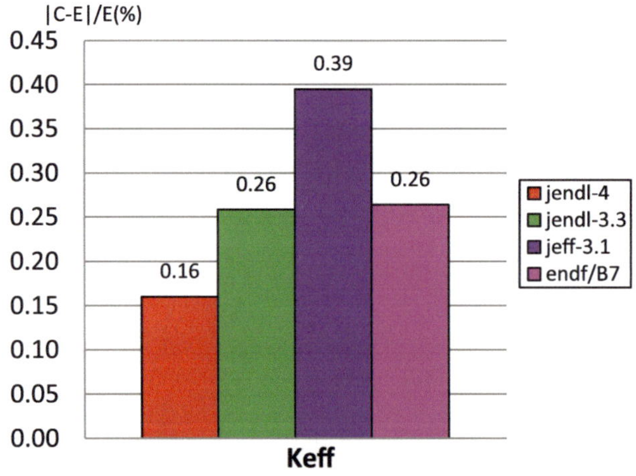


**Figure 4.**
*The |C-E|/E ratio of $k_{eff}$ for the fast benchmarks in $^{233}U$.*

From **Figures 3** and **4** we find that the best results of criticality are given by JENDL-4 with a deviation from the experience of 0.16%, secondly, we find ENDF/B-VII with a deviation by compared to the experience of 0.26% we note an improvement during the transition from JENDL-3.3 to JENDL-4. We also notice an overestimation of JEFF-3.1 of the criticality with a deviation from the experience of 0.39%.

## 5.2 Fast benchmarks in Pu-239

**Figures 5** and **6** represent the variation of $k_{eff}$ according to the cases for the fast benchmarks in $^{239}Pu$ isotope as well as the average deviations from the experience.

In **Figures 5** and **6**, the variation of $k_{eff}$, shows that the process based on ENDF/B-VII and JEFF-3.1 gives results that are in good agreement with the experimental values, the deviations from the experiment are 0.34% and 0.33% respectively, with

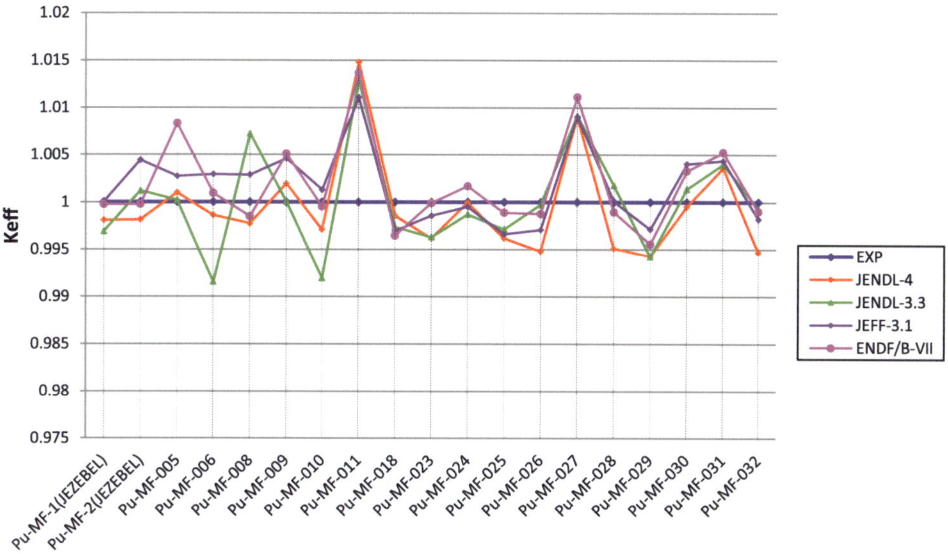


**Figure 5.**
*$k_{eff}$ depending on the case for the fast benchmarks in Pu-239.*

the exception of JENDL-3.3 and JENDL-4 which have deviations from criticality greater than the other libraries of 0.39% and 0.38% respectively. Apparently, JENDL-3.3 gives $k_{eff}$s far from 1 compared to other libraries in the Pu-MF-006 and Pu-MF-008 and Pu-MF-010 benchmarks as we notice that the problem is corrected in JENDL-4, and JENDL-4 far from 1 compared to other libraries in benchmarks Pu-MF-026, Pu-MF-28 and Pu-MF-32, We note that there is a deterioration during the transition from JENDL-3.3 to JENDL-4 in these three benchmarks. At the three benchmarks Pu-MF-11, PU-MF-27 and PU-MF-31 all the libraries have criticality estimates.

The Pu-MF-26, 28 and 32 benchmarks use stainless steel as a reflector, so the JENDL-4's underestimation of criticality compared to other libraries and due to the overestimation of JENDL-4 to other libraries in the cross-section of carbon capture.

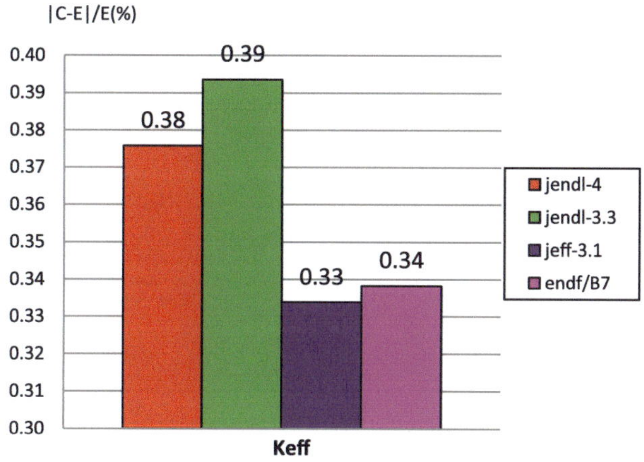


**Figure 6.**
*The |C-E|/E ratio of $k_{eff}$ in the case of fast benchmarks in Pu-239.*

## 6. Conclusions

In this work we were able to model rapid critical benchmarks using the main fissile nuclei which are, the $^{235}$U, the $^{233}$U, and the $^{239}$Pu, we previously generated cross-sections using the NJOY code, these cross-sections come from the main evaluations ENDF/B-VII, JEFF-3.1, JENDL-3.3 and JENDL-4.

The Monte Carlo calculation that we carried out consisted in determining the $k_{eff}$ parameter, the difference between the calculation and the experiment depends mainly on the type of evaluation used as well as the fissile core of the benchmarks considered this difference remains acceptable all the same. So that we can say, that our results are in good agreement with those obtained experimentally.

## Author details

Sanae El Ouahdani[1*], Hamid Boukhal[2], El Mahjoub Chakir[3], Ahmed Gaga[1], Houda Elyaakoubi[2], Mustapha Makhloul[2], Abdelaziz Ahmed[4], Abdessamad Didi[5] and Mohamed Bencheikh[6]

1 Polydisciplinary Faculty, LRPSI Laboratory, Physics Department, Sultan Moulay Slimane University, Beni Mellal, Morocco

2 Faculty of Sciences, ERSN, Abdelmalek Essaadi University, Tetouan, Morocco

3 Faculty of Sciences, LHESIR, Ibn Tofail University, Kenitra, Morocco

4 Faculty of Lawder, Physics Department, University of Abyan, Abyan, Yemen

5 National Center for Energy Sciences and Nuclear Techniques, Rabat, Morocco

6 Faculty of Sciences and Technologies, Physics Department, Mohammedia Hassan II University of Casablanca, Mohammedia, Morocco

*Address all correspondence to: selouahdani@gmail.com

## References

[1] Chadwick MB et al. Nuclear data sheets. 2006;**107**(12):2931-3060

[2] The JEFF-3.1 Nuclear Data Library. JEFF Report 21. ISBN 92-64-02314-3

[3] Shibata K et al. Japanese evaluated nuclear data library version 3 revision-3: JENDL-3.3. Journal of Nuclear Science and Technology. 2002;**39**:1125

[4] Shibata K et al. JENDL-4.0: A new library for nuclear science and engineering. Journal of Nuclear Science and Technology. 2011;**48**(1):1-30

[5] NJOY 99.9. A Code System for Producing Pontwise and Multigroup Neutron and Photon Cross Sections for ENDF/B, Evaluated Nuclear Data. New Mexico: Los Alamos National Laboratory. 1999

[6] International Handbook of Evaluated Criticality Safety Benchmark Experiments. NEA/NSC/DOC (95). Vol. I-VII. 2003

[7] MCNPXTM user's manual. Version 2.7.0. LA-CP-11-00438. Manuel/Los Alamos National Laboratory; 2011

[8] JANIS 3.0 user's guide. 2007

Chapter 6

# Applications of Fuzzy Set and Fixed Point Theory in Dynamical Systems

*Praveen Kumar Sharma, Shivram Sharma, Jitendra Kaushik and Palash Goyal*

## Abstract

This chapter shall discuss various applications of fixed-point theory and fuzzy set theory. Fixed point theory and fuzzy set theory are very useful tools that are applicable in almost all branches of mathematical analysis. There are many problems that cannot be solved by applying the concept of other existing theories but can be solved easily by using the concept of fuzzy set theory and fixed point theory. So here in this chapter, we shall introduce the fuzzy set theory and fixed point theory concerning their applications in existing branches of science, engineering, mathematics, and dynamical systems.

**Keywords:** fixed point, fuzzy set, dynamical systems, stability, fuzzy differential equations, integral equations

## 1. Introduction

Fixed point theory is an area of mathematics linked to functional analysis and topology that is still in its infancy. Fixed point theory is an important subject in the fast-growing domains of nonlinear analysis and nonlinear operators. It is a relatively new scientific area that is developing rapidly. In topics as diverse as differential equations, topology, economics, game theory, dynamics, optimal control, and functional analysis, fixed points and fixed point theorems have always been important theoretical tools. Furthermore, with the development of accurate and efficient techniques for computing fixed points, the concept's relevance for applications has expanded dramatically, making fixed point methods a vital weapon in the arsenal of the applied mathematician.

Set theory, general topology, algebraic topology, and functional analysis are just a few of the major fields of mathematics that give natural settings for fixed point theorems. Approximation theory, potential theory, game theory, mathematical economics, theory of differential equations, and other disciplines use fixed point theorems to solve problems in approximation theory, potential theory, game theory, mathematical economics, and so on. It is possible to evaluate various problems from science and engineering using fixed point approaches when one is concerned with a system of differential/integral/functional equations. This method is particularly beneficial when dealing with control system issues and the idea of elasticity.

Fixed point theorems are the most important tools for proving the existence and uniqueness of solutions to various mathematical models (differential, integral, and partial differential equations, variational inequalities, and so on), which represent phenomena arising in multiple fields such as steady-state temperature distributions, chemical reactions, Neutron transport theories, economic theories, epidemics, and fluid flow. They are also employed to look at the difficulty of determining the best central for these systems.

Let $F : X \to X$ represent a function on the set $X$. A point $x \in X$ is called a fixed point of $F$ if $F(x) = x$, that is, a point, which remains invariant under the transformation $F$, is called a fixed point, and fixed point theorems are theorems that deal with the attributes and existence of fixed points. If $F$ is a function defined on the real numbers as $F(x) = x + 2$, then it has no fixed points since $x$ is never equal to $x + 2$ for any real number.

Let $F : [0,1] \to [0,1]$ be defined by $x/10$ then $F\ (0) = 0$. Hence 0 is a fixed point of $F$.

Poincare [1] was the first to establish the fixed-point theory in 1886. He arrived at the first result on a fixed point using a continuous function.

Browder [2] proved the following useful theorem in 1912. Browder's fixed point theorems are fundamental in fixed point theory and its applications. Browder's fixed-point theorem states, "If $C$ is a unit ball in $E^n$ (Euclidean n-dimensional space) and $T : C \to C$ a continuous function. Then $T$ has a fixed point in $C$, or $Tx = x$ has a solution."

The Particular Case of this theorem on the real line can be stated in the following way.

Let $T : [0,1] \to [0,1]$ be a continuous function. Then $T$ has a fixed point.

Schauder proved the following theorem for compact maps.

Let $X$ be a Banach space and let $C$ be a closed, bounded subset of $X$. Let $T : C \to C$ be a compact map. Then $T$ has at least one fixed point in $C$.

This theorem is important in the numerical treatment of equations in analysis.

Banach [3] investigated the concept of contraction type mappings in metric space in 1922. Using the condition of contraction mapping, he established an interesting conclusion in metric space. "Every contraction mapping of a complete metric space into itself has a unique fixed point," according to the Banach contraction principle.

A contraction mapping is continuous, but a continuous map is not necessarily a contraction.

For example, a translation map $T : R \to R$ is defined by $Tx = x + p, p > 0$, is continuous but not contraction.

The Banach contraction principle has a lot of uses, but it has one major flaw: It requires the function to be consistent throughout the space. Kannan [4] proved the improved conclusion in fixed point theory to avoid this flaw. The Banach contraction principle has a lot of uses, but it has one major flaw: It requires the function to be consistent throughout the space. Kannan [5] proved the improved conclusion in fixed point theory to avoid this flaw. He proved that "let $F$ be a self-mapping of complete metric space $X$ satisfying the following inequality

$$d(Fx, Fy) \leq \alpha\ [d\ (x, Fx) + d\ (y, Fy)]\ for\ all\ x, y \in X, 0 < \alpha\ <\ 1/2.$$

Then $F$ has a unique fixed point."

Jungck [6] generalized Banach's fixed point theorem by proving a common fixed point theorem for commuting maps. The Banach fixed point theorem has many applications, but it has one flaw: The definition necessitates function continuity.

In 1982, Sessa [7] refined Jungck's result and proposed the concept of weakly commuting mappings in metric space, demonstrating that "two commuting mappings also commute weakly, but two weakly commuting mappings are not certainly commuting."

Jungck [8] achieved a breakthrough when he declared the new concept of "compatibility" of mappings and demonstrated its utility in achieving a common fixed point of mappings. Every weak commutative pair of mappings is compatible, according to Jungck [9], but the opposite does not have to be true. In his study [10], Singh points out that commutativity does not entail the presence of a series of points that satisfy the compatibility criterion.

Jungck et al. [11] introduced the concept of compatible mappings of type (A) in 1993 and proved some common fixed point theorems.

Common fixed-point theorems for Mann-type iterations are useful for common fixed point theorems and their applications to the best approximation.

In 1999, Popa [12] proved some fixed point theorems for compatible mappings satisfying an implicit relation. Some other results of Popa have been targeted by many authors and common fixed point theorems in different spaces on using implicit relations.

The notion of convex metric spaces was initially introduced by Takahashi [13]. He and others gave some fixed point theorems for non-expansive mappings in convex metric spaces.

In the metric space setting, the strict contractive condition does not ensure the existence of a common fixed point unless the space is assumed compact or the tough conditions are replaced by stronger conditions as in [14, 15]. In 1986, Jungck [8] introduced the notion of compatible mappings. This concept was frequently used to prove the existence of theorems in common fixed point theory. However, the study of common fixed points of non-compatible mappings is also very interesting.

In an attempt to study fixed points of non-self-mappings, Assad and Kirk [16] gave sufficient conditions for such mappings to have a fixed point by proving a result for multivalued mappings in convex metric spaces. Naimpally et al. [17] proved fixed point theorems in convex metric spaces.

Gähler [18] introduced the concept of 2-metric space and further studied 2-metric and other spaces in [19, 20]. He defined 2-metric space as a real-valued function of a point triples on a set X, whose abstract properties were suggested by the area function in Euclidean space. It is natural to expect 3-metric space, which is indicated by the volume function.

In 1998, Pant [21] introduced the notion of R-weakly commuting maps and point-wise R-weakly commuting maps in metric spaces. He has observed that two self-maps on a metric space can fail to be point-wise R-weakly commuting only if they possess a coincidence point at which they do not commute.

The systematic study of fixed points of multivalued mappings had been started with the work of Nadler [22] in 1969, who proved that any multivalued contractive mapping of a complete metric space X into the family of a closed bounded subset of X has a fixed point. Ciric [23] was the first to prove the most general fixed point theorem for a generalized multivalued contraction mapping.

Naimpally et al. [1] obtained some interesting results on fixed point and coincidence point theorems for a hybrid of multivalued and single-valued maps satisfying a contraction condition.

In 1942, Menger [24] suggested associating a distribution function in place of a distance function to any two points in metric space and introduced the concept of probabilistic metric space under statistical metric space.

Sehgal [25] initiated the study of contraction mapping on probabilistic metric space in 1966. Sehgal and Bharucha-Reid [26] proved the Banach contraction principle for probabilistic metric space, stating that "A contraction mapping on a complete probabilistic metric space has a unique fixed point."

Due to the various paradigmatic changes in science and mathematics, many changes also took place in the concept of uncertainty. One such change is the concept of uncertainty which is at the stage of transition from the traditional view to the modern view and is characterized chiefly by the theories of the uncertainty of probability theory.

An important point in the evolution of the modern concept of uncertainty was the publication of a seminal paper by Zadeh [27], a computer scientist university of California; the U.S.A. was the first who introduce the concept of fuzzy set theory in his seminal paper as a new way to represent vagueness in our everyday life. Zadeh introduced a theory whose objects fuzzy sets are sets with boundaries that are not precise. The membership in a fuzzy set is not a matter of affirmation or denial but rather a degree. This concept is being used and found to be more appropriate in solving problems of all disciplines.

The concept of fuzzy sets was initially introduced by Zadeh [27] in 1965 and has caused great interest among pure and applied mathematicians. It has also raised enthusiasm among engineers, biologists, psychologists, and economists.

Let X be a set; we define a fuzzy set X as a map $M : X \to I = [0, 1]$. It is to be remarked that fuzzy sets can be regarded as a generalization of characteristic functions taking values between 0 and 1 (including 0 and 1), and the characteristic function on a set $X$ is the constant mapping.

In classical set theory, a subset $A$ of a set $X$ can be defined by its characteristic function $\chi_A(x) = 0, \textit{if } x \notin A$ and $\chi_A(x) = 1, \text{if } x \in A$.

The mapping may be represented as a set of ordered pairs $\{(x, \chi_A(x))\}$ with exactly one ordered pair present for each element of $X$. The first element of the ordered pair is an element of the set $X$, and the second is its value $\{0, 1\}$. The value 0 is used to represent non-membership, and the value 1 is used to describe the membership of the element $A$. The truth or falsity of the statement, $x$ is in $A$, determined by the ordered pair. The statement is true if the second element of the ordered pair is 1, and the statement is false if it is 0. Similarly, a fuzzy subset $A$ of a set $X$ can be defined as a set of ordered pairs $\{(x, \chi_A(x)) : x \in X\}$, each with the first element from $X$ and the second element from the interval $[0, 1]$ with exactly one ordered pair present for each component of $X$. This defines a mapping $\mu_A$ between elements of the set X and the values in the interval $[0, 1]$. That is, $\mu_A : X \to [0, 1]$.

The value 0 represents complete non-membership, the value 1 represents complete membership, and the values in between represent intermediate degrees of membership.

The fuzzy set theory has an application in neural network theory, robotics, reliability, stability theory, mathematical programming, modeling theory, engineering sciences, medical sciences, image processing, control theory, communication, etc.

In the last three decades, fuzzy mathematics's development and rich growth were tremendous. A large number of authors studied applications of fuzzy set theory in different engineering branches.

Zimmermann and Sebastian [28, 29] defined knowledge base system design and intelligent system design support. Tsourveloudis et al. [30] defined machine flexibility and established a result.

For fuzzy mathematics, we refer to Bedard [31], Butnariu [32], Grabiec [33], and Weiss [34].

## 2. Main results/discussion/application of fuzzy set and a fixed point in a dynamical system

In this chapter, our main aim is to give an application of fuzzy sets and fixed points in a dynamic system.

**A dynamical system** is one in which a function explains the time dependency of a point in an ambient space or one in which something evolves with time. For instance, mathematical models that represent the swinging of a clock pendulum, water flow in a conduit, the quantity of fish in a lake each spring, population expansion, and so on.

A dynamic system can be described over either discrete time steps or a continuous timeline.

**Discrete-time dynamical system-**

$$x_t = F(x_{t-1}, t) \tag{1}$$

This type of model is called a difference equation, a recurrence equation, or an iterative map (if the right-hand side is not dependent on $t$)

**Continuous-time dynamical system-**

$$\frac{dx}{dt} = F(x, t) \tag{2}$$

This type of model is called a differential equation.

In both cases, $x_t$ or $x$ is the system's state variable at time $t$, which may take a scalar or vector value. $F$ is a function that determines the rule by which the system changes its state over time.

Dynamical systems are often modeled by differential equations.

So, here we discuss an application of fuzzy Laplace transforms to solve differential equations/fuzzy differential equations:

Fuzzy differential equations (FDEs) are a natural technique to model dynamic systems under uncertainty. One of the most basic FDEs, first-order linear fuzzy differential equations, can be found in many applications. Chang and Zadeh [35] were the first to present the fuzzy derivative notion. The concept of FDEs was used in the analysis of fuzzy dynamical problems by Kandel and Byatt [4, 36]. FDEs and their fuzzy beginning and boundary value problems are solved using the fuzzy Laplace transform approach. Fuzzy Laplace transforms make it easier to solve an FDE by converting it to an algebraic problem.

Operational calculus is an important area of applied mathematics that involves switching from calculus operations to algebraic operations on transforms. The fuzzy

Laplace transform approach is practically the essential functional method for engineers. The fuzzy Laplace transform also benefits by directly addressing difficulties, fuzzy initial value problems without identifying a general solution, and non-homogeneous differential equations without first solving the corresponding homogeneous equation.

There are a number of mathematicians who studied and developed several approaches to study FIVP [37–41]. They initially used H-Differentiability for fuzzy-valued functions. Using this concept, they looked at the existence and uniqueness of the solution of FIVP [37, 41, 42]. This concept has a drawback: The fuzzy solution behaves quite differently from the crisp solution. Bede B and Gal SG [43] introduced a new idea called the strongly generalized differentiability, and it was studied and used for solving FIVPs in [44–47]. This concept allows us to overcome the drawback mentioned above. So, we use this differentiability concept to find out the solution to FIVP in this chapter.

The fuzzy Laplace transform method solves the problems of FDEs and corresponding fuzzy initial and boundary values. This method solves FIVPs/FDEs directly and gives the complete solution without determining the complementary and particular solution (one can refer to [44, 45, 48–51]. This chapter uses this technique to solve FIVPs/FDEs/ODEs.

We need some definitions and theorems given under the following to solve the fuzzy differential equation by the fuzzy Laplace transform.

**Definition 2.1.** ([46]). Let $f(x)$ be a continuous fuzzy-value function. Suppose that $f(x)\odot e^{-px}$ is improper fuzzy Riemann integrable on $[0,\infty)$, then $\int_0^\infty f(x)\odot e^{-px}dx$ is called fuzzy Laplace transforms and is denoted as

$$L[f(x)] = \int_0^\infty f(x)\odot e^{-px}dx = \left(\int_0^\infty \underline{f}(x,r)e^{-px}dx, \int_0^\infty \overline{f}(x,\alpha)e^{-px}dx\right)$$
$$= \left(l\left[\underline{f}(x,\alpha)\right], l\left[\overline{f}(x,\alpha)\right]\right),$$

where $\left[\underline{f}(x,\alpha)\right] = \int_0^\infty \underline{f}(x,\alpha)e^{-px}dx$ , $l\left[\overline{f}(x,\alpha)\right] = \int_0^\infty \overline{f}(x,\alpha)e^{-px}dx$

(Or) $L[f(x)] = \left(l\left[\underline{f}(x,\alpha)\right], l\left[\overline{f}(x,\alpha)\right]\right)$

**Theorem 2.2.** Chalco-Cano et al. [46] Let $f : R \to E$ be a fuzzy valued function and denote $f(t) = \left(\underline{f}(t,\alpha), \overline{f}(t,\alpha)\right)$, for each $\alpha \in [0,1]$. Then

1. If $f$ is (i)-differentiable, then $\underline{f}(t,\alpha)$ and $\overline{f}(t,\alpha)$ are differentiable functions and $f'(t) = \left(\underline{f}'(t,\alpha), \overline{f}'(t,\alpha)\right)$

2. If $f$ is (ii)-differentiable, then $\underline{f}(t,\alpha)$ and $\overline{f}(t,\alpha)$ are differentiable functions and $f'(t) = \left(\overline{f}'(t,\alpha), \underline{f}'(t,\alpha)\right)$

**Formulae 2.3.** Consider the fuzzy initial value problem

$$\begin{cases} y'(t) = f(t, y(t)) \\ y(0) = \left(\underline{y}(0,\alpha), \overline{y}(0,\alpha)\right), 0 < \alpha \le 1. \end{cases}$$

Where $f: R_+ \times E \to E$ is a continuous fuzzy mapping. Using the fuzzy Laplace transform method, we have: $L[y'(t)] = L[f(t, y(t))]$

Case I—If we consider $y'(t)$ by using (i)-differentiable, then $y'(t) = \left(\underline{y}'(t,\alpha), \bar{y}'(t,\alpha)\right)$ and

$$L[y'(t)] = (s \odot L[y(t)]) -^h y(0)$$

(Or) $l\left[\underline{f}(t, y(t), \alpha)\right] = s\, l\left[\underline{y}(t,\alpha)\right] - \underline{y}(0,\alpha),\, l\left[\bar{f}(t, y(t), \alpha)\right] = s\, l[\bar{y}(t,\alpha)] - \bar{y}(0,\alpha)$

(Or) $l\left[\underline{f}(t, y(t), \alpha)\right] = s\, H_1(s,\alpha) - \underline{y}(0,\alpha),\, l\left[\bar{f}(t, y(t), \alpha)\right] = s\, K_1(s,\alpha) - \bar{y}(0,\alpha)$

Where $l\left[\underline{y}(t,\alpha)\right] = H_1(s,\alpha)$ , $l[\bar{y}(t,\alpha)] = K_1(s,\alpha)$

Case II—If we consider $y'(t)$ by using (ii)-differentiable, then $y'(t) = \left(\bar{y}'(t,\alpha), \underline{y}'(t,\alpha)\right)$ and

$$L[y'(t)] = y(0) -^h (-s \odot L[y(t)])$$

(Or) $l\left[\underline{f}(t, y(t), \alpha)\right] = s\, l\left[\underline{y}(t,\alpha)\right] - \underline{y}(0,\alpha),\, l\left[\bar{f}(t, y(t), \alpha)\right] = s\, l[\bar{y}(t,\alpha)] - \bar{y}(0,\alpha)$

(Or) $l\left[\underline{f}(t, y(t), \alpha)\right] = s\, H_1(s,\alpha) - \underline{y}(0,\alpha),\, l\left[\bar{f}(t, y(t), \alpha)\right] = s\, K_1(s,\alpha) - \bar{y}(0,\alpha)$

Where $l\left[\underline{y}(t,\alpha)\right] = H_2(s,\alpha)$ , $l[\bar{y}(t,\alpha)] = K_2(s,\alpha)$

Now, we solve the fuzzy differential equation by the fuzzy Laplace transform method-

**Example 2.4.** Consider the initial value problem

$$\begin{cases} y'(t) = y(t) 0 \le t \le T \\ y(0) = \left(\underline{y}(0,\alpha), \bar{y}(0,\alpha)\right). \end{cases}$$

by using the fuzzy Laplace transform method, we have

$L[y'(t)] = L\,[y(t)]$, and $L[y'(t)] = \int_0^\infty y'(t) \odot e^{-pt} dt$ in (i)-differentiable then by using Case (i), we have $L[y'(t)] = (s\, L\,[y(t)]) \ominus y(0)$

Therefore, $L[y(t)] = s\, L\,[y(t)]) \ominus y(0)$

$$\begin{aligned} l[\bar{y}(t,\alpha)] &= s\, l\left[\underline{y}(t,\alpha)\right] - \underline{y}(0,\alpha) \\ l\left[\underline{y}(t,\alpha)\right] &= s\, l[\bar{y}(t,\alpha)] - \bar{y}(0,\alpha) \quad \ldots \} \end{aligned} \tag{3}$$

Hence, the solution of system (3) is:

$$l[\bar{y}(t,\alpha)] = -\bar{y}(0,\alpha)\left(\frac{s}{s^2-1}\right) + \underline{y}(0,\alpha)\left(-\frac{1}{s^2-1}\right)$$

$$l\left[\underline{y}(t,\alpha)\right] = -\underline{y}(0,\alpha)\left(\frac{s}{s^2-1}\right) + \bar{y}(0,\alpha)\left(-\frac{1}{s^2-1}\right)$$

Thus

$$\bar{y}(t,\alpha) = -\bar{y}(0,\alpha)l^{-1}\left[\left(\frac{s}{s^2-1}\right)\right] + \underline{y}(0,\alpha)l^{-1}\left[\left(-\frac{1}{s^2-1}\right)\right]$$

$$\underline{y}(t,\alpha) = -\underline{y}(0,\alpha)\,l^{-1}\left[\left(\frac{s}{s^2-1}\right)\right] + \bar{y}(0,\alpha)\,l^{-1}\left[\left(-\frac{1}{s^2-1}\right)\right]$$

Finally, we have:

$$\bar{y}(t,\alpha) = e^{-t}\left(\frac{\underline{y}(0,\alpha)-\bar{y}(0,\alpha)}{2}\right) - e^{t}\left(\frac{\underline{y}(0,\alpha)+\bar{y}(0,\alpha)}{2}\right)$$

$$\underline{y}(t,r) = e^{-t}\left(\frac{-\underline{y}(0,\alpha)+\bar{y}(0,\alpha)}{2}\right) - e^{t}\left(\frac{\underline{y}(0,\alpha)+\bar{y}(0,\alpha)}{2}\right)$$

If $y'(t)$ in (ii)-differentiable, then by using Case II, we have $L[y'(t)] = -(y(0))\ominus(-s\,L\,[y(t)])$. Therefore, $L[y(t)] = (-y(0))\ominus(-s\,L\,[y(t)])$

$$\left.\begin{aligned} l[\bar{y}(t,\alpha)] &= s\,l[\bar{y}(t,\alpha)] - \bar{y}(0,\alpha) \\ l\left[\underline{y}(t,\alpha)\right] &= sl\left[\underline{y}(t,\alpha)\right] - \underline{y}(0,\alpha) \quad ... \end{aligned}\right\} \tag{4}$$

Hence, the solution of system (4) is:

$$l[\bar{y}(t,\alpha)] = -\bar{y}(0,\alpha)\left(\frac{1}{1+s}\right)$$

$$l\left[\underline{y}(t,\alpha)\right] = -\underline{y}(0,\alpha)\left(\frac{1}{1+s}\right)$$

Thus

$$\bar{y}(t,\alpha) = -\bar{y}(0,\alpha)l^{-1}\left(\frac{1}{1+s}\right)$$

$$\underline{y}(t,\alpha) = -\underline{y}(0,\alpha)l^{-1}\left(\frac{1}{1+s}\right)$$

Finally, we have:

$$\bar{y}(t,\alpha) = -\bar{y}(0,\alpha)e^{-t}$$

$$\underline{y}(t,\alpha) = -\underline{y}(0,\alpha)e^{-t}$$

**Remark 2.5**. By following the above procedure, the solution of fuzzy IVP with initial conditions can be obtained. The solution of simultaneous fuzzy linear differential equations and fuzzy BVPs can also be obtained.

Now, we discuss the application of fixed points in existence and the uniqueness of the solution of the ordinary differential equation.

IVP's existence and uniqueness can be easily established using the fixed point technique. Banach fixed point theorem can be applied to derive the existence and uniqueness of the solution of an initial value problem. The function used in IVP satisfies the Lipschitz condition.

**Definition 2.6**. Contraction Mapping:

Let X be a complete normed linear space (Banach Space). A mapping $F : X \to X$ is called a contraction if $\|Fx - Fy\| \leq \alpha \|x - y\|, \forall x, y \in X$, for some $\alpha < 1$

Example-. If $F(x) = \frac{x}{2}$, then $F$ is a contraction for $\alpha = \frac{1}{2}$

**Definition 2.7**. Banach Fixed-Point Theorem (or Banach Contraction Principle):

If $F : X \to X$ is a contraction, then $F$ has a unique fixed point; say it is $x_1$. $Fx_1 = x_1$

Further, the sequence $\{x_n\}$ defined by $x_n = Fx_n$ $(\forall n = 1, 2, 3, \ldots)$, converges to the unique fixed point $x_1$ of $F$.

**Definition 2.8**. Generalized Banach contraction principle:

If $F^n$ is contraction, $F : X \to X$ is a Banach space for $n \geq 1$, then $F$ has a unique fixed point

**Theorem 2.9.** Consider an Initial Value Problem (IVP)

$$\frac{dy}{dx} = f(x, y), y(x_0) = y_0$$

Let $f(x, y)$ be a continuous function defined on a domain $D \subseteq R^2$. Let f be Lipschitz continuous concerning y on D. Then, there exists a unique solution to the IVP on an interval $|x - x_0| \leq h$, where $h = \min\left(a, \frac{b}{M}\right), M = Max\, |f(x, y)|$

$(x, y) \in R$, and $R = \left\{(x, y) : |x - x_0| \leq a, \left|y - y_0\right| \leq b \right\} \subset D$

Further, the unique solution can be computed from the successive approximation scheme $y_{n+1}(x) = y_0 + \int_{x_0}^{x} f\left(t, y_n(t)\right) dt, y_0(t) = y_0$ $(\forall n = 0, 1, 2, \ldots)$

**Proof:** The solvability of IVP follows

If the integral equation $y(x) = y_0 + \int_{x_0}^{x} f(t, y(t)) dt$ is solvable, let $X = C[x_0, x_1]$ set of all continuous functions defined on $[x_0, x_1]$. Define $\|x\| = \sup_{t \in [x_0, x_1]} |x(t)|$ $(X, \|.\|)$ is complete normed linear space.

Define

$$y(x) = y_0 + \int_{x_0}^{x} f(t, y(t)) dt \tag{5}$$

Define an operator $F : C[x_0, x_1] \to C[x_0, x_1]$ by $(Fy)(x) = y_0 + \int_{x_0}^{x} f(t, y(t)) dt$.

If $F$ has a fixed point, that is, there exists a $y$ such that $y = Fy$, then the fixed point $y$ is a solution to the integral equation (5).

Now, we will show $F^n$ Is contraction for some large $n$.

$$(Fy)(x) = y_0 + \int_{x_0}^{x} f(t, y(t)) dt$$

Let $y_1, y_2 \in C[x_0, x_1]$, then

$$\left|(Fy_1)(x) - (Fy_2)(x)\right| = \left|\int_{x_0}^{x} (f(t, y_1(t) - f(t, y_2(t)) dt\right|$$

$$\leq \alpha \int_{x_0}^{x} \left|y_1(t) - y_2(t)\right| dt \tag{6}$$

$$\leq \alpha \int_{x_0}^{x} \sup_{t \in [x_0, x_1]} |y_1(t) - y_2(t)| dt$$

$$\leq \alpha \int_{x_0}^{x} \| y_1(t) - y_2(t) \| dt$$

$$\leq \alpha (x - x_0) \| y_1 - y_2 \| \tag{7}$$

$$\left| (F^2 y_1)(x) - (F^2 y_2)(x) \right| = \left| F(Fy_1)(x) - F(Fy_2)(x) \right|$$

$$\leq \alpha \int_{x_0}^{x} |Fy_1(t) - Fy_2(t)| dt$$

$$\leq \alpha^2 \int_{x_0}^{x} (t - x_0) \| y_1 - y_2 \| dt$$

$$\leq \frac{\alpha^2}{2} (x - x_0)^2 \| y_1 - y_2 \|$$

On taking sup, we have

$$\| (F^2 y_1) - (F^2 y_2) \| \leq \frac{\alpha^2}{2!} (x_1 - x_0)^2 \| y_1 - y_2 \|$$

Similarly,

$$\| (F^3 y_1) - (F^3 y_2) \| \leq \frac{\alpha^3}{3!} (x_1 - x_0)^3 \| y_1 - y_2 \|$$

$$- - -$$

$$\| (F^n y_1) - (F^n y_2) \| \leq \frac{\alpha^n}{n!} (x_1 - x_0)^n \| y_1 - y_2 \| \tag{8}$$

In (8), $\frac{\alpha^n}{n!}(x_1 - x_0)^n < 1$ if $n$ is large enough

This implies that $F^n$ It is a contraction for large $n$, and $F$ has a unique fixed point.

This implies that there is a unique solution to the integral equation.

This shows that there is a unique solution to IVP.

Further, $x_{n+1} = Fx_n$, $\{x_n\}$ converges to the unique solution of the IVP.

This implies that $y_{n+1} = Fy_n$

Which implies that $y_{n+1}(x) = y_0 + \int_{x_0}^{x} f(t, y_n(t)) dt, y_0(x) = y_0$

**Example 2.10**. Consider the IVP $\frac{dy}{dx} = \frac{2y}{x} y(1) = 1$

Here $f(x, y) = \frac{2y}{x}, x_0 = 1, y_0 = 1$

$$y_{n+1}(x) = y_0 + \int_{x_0}^{x} f(t, y_n(t)) dt$$

$$y_1(x) = 1 + \int_{1}^{x} f(t, y_0(t)) dt$$

$$y_1(x) = \int_{1}^{x} \frac{2}{t} dt$$

$$y_1(x) = 2 \log(x)$$

$$y_2(x) = 1 + \int_1^x f(t, y_1(t))dt$$

$$y_2(x) = \int_1^x 4\,\frac{logt}{t}dt$$

$$y_2(x) = \int_1^x 4\frac{logt}{t}dt$$

$$y_2(x) = 2(x^2 - 1)$$

$$y_3(x) = \int_1^x f(t, y_2(t))dt$$

$$y_3(x) = 1 + \int_1^x f(t, y_2(t))dt$$

$$y_3(x) = 1 + \int_1^x \frac{4(t^2 - 1)}{t}dt$$

$$y_3(x) = 1 + 4\int_1^x \left(t - \frac{1}{t}\right)dt$$

$$y_3(x) = 1 + 2\,(x^2 - 1) - 4\,(\log x)$$

$$y_3(x) = (2x^2 - 1) - 4\,(\log x)$$

On keep continuing this process up to the nth time and then taking $n \to \infty$ we have

$$y(x) = x^2$$

**Example 2.11**. Consider the IVP $\frac{dy}{dx} = y, y(0) = 1$
Here $f(x, y) = y, x_0 = 0, y_0 = 1$

$$y_{n+1}(x) = y_0 + \int_{x_0}^x f(t, y_n(t))dt$$

$$y_1(x) = 1 + \int_0^x y_0(t)dt$$

$$y_1(x) = 1 + \int_0^x 1dt = 1 + x$$

$$y_2(x) = 1 + \int_0^x y_1(t)dt = 1 + x + \frac{x^2}{2}$$

$$y_3(x) = 1 + \int_0^x y_2(t)dt = 1 + x + \frac{x^2}{2} + \frac{x^3}{6}$$

$$- - -$$

$$y_n(x) = 1 + x + \frac{x^2}{2!} + \frac{x^3}{3!} + \ \ldots\ + \frac{x^n}{n!} = \sum \frac{x^n}{n!}$$

Taking $n \to \infty$ , then we have $y_n(x) \to e^x$
Thus $y(x) = e^x$ Is a solution for a given IVP.

Stability of the solution: In these questions/examples, the functions $f(x,y) = \frac{2y}{x}$ and $f(x,y) = y$ satisfies the Lipschitz condition concerning the initial condition $y_0 = 1$, so the answer is stable concerning initial data.

## 3. Conclusion

Here in this chapter, we discussed the application of a fuzzy set and a fixed point in dynamic systems. We also tried to discuss applications of fuzzy sets and fixed points in other directions. We gave a solution of fuzzy ordinary differential equations with initial conditions by the fuzzy Laplace transform method and provided a solution to the existence and uniqueness problem of ODE of the first order by the fixed point technique. We solved examples of existence and uniqueness problems and checked the stability of the solution also.

## Author details

Praveen Kumar Sharma[1*], Shivram Sharma[2], Jitendra Kaushik[3] and Palash Goyal[4]

1 Department of Mathematics, SVIS, Shri Vaishnav Vidyapeeth Vishwavidyalaya, Indore, M.P., India

2 Department of Mathematics, Govt. P.G. College, Guna, M.P., India

3 Chandigarh University, Mohali, Punjab, India

4 University Institute of Technology, RGPV, Shivpuri, M.P., India

*Address all correspondence to: praveen_jan1980@rediffmail.com

## References

[1] Poincare H. Analysis situs. J. det. Ecole Polytrch. 1985;**2**:1-123

[2] Browder FE. On a generalisation of Schauder's fixed point theorem. Duke Mathematical Journal. 1959;**26**: 291-303

[3] Banach S. Theorie Les Operations Lineaires, Manograie Mathematyezne Warsaw Poland, In French Z. Subwencji Funduszu Kultury Narodowej, New York. Warsaw: Z Subwencji Funduszu Kultury Naroldowej; 1932

[4] Kandel A, Byatt WJ. Fuzzy differential equations, Proceedings of International Conference on Cybernetics and Society. Tokyo, Japan: IEEE; 1978, pp. 1213-1216

[5] Kannan R. Some results on fixed points II. American Mathematical Monthly. 1969;**76**:405-408

[6] Jungck G. Commuting mappings and fixed points. American Mathematically Monthly. 1976;**83**:261-263

[7] Sessa S. On weak commutativity condition of mappings in consideration of a fixed point. Publications de l'Institut Mathematique. 1982;**32**(46): 149-153

[8] Jungck G. Compatible mappings and common fixed points. International Journal of Mathematics and Mathematical Sciences. 1986;**9**:771-779

[9] Jungck G. Compatible mappings and common fixed points. International Journal of Mathematics and Mathematical Sciences. 1988;**11**:285-288

[10] Sherwood H. Complete probabilistic metric spaces. Z. Wahrseh Verw. Gabinete. 1971;**20**:117-228

[11] Jungck G, Murthy PP, Cho YJ. Compatible mappings of type (A) and common fixed points. Mathematica Japonica. 1993;**38**(2):381-390

[12] Popa V. Some fixed point theorems for compatible mappings satisfying an implicit relation. Demonstratio Mathematica. 1999;**32**(1):157-163

[13] Takahashi W. A convexity in metric spaces and non-expansive mappings. Kodai Mathematical Semester Report. 1970;**22**:142-149

[14] Jachymski J. Common fixed-point theorems for some maps families. Indian Journal of Pure and Applied Mathematics. 1994;**25**:925-937

[15] Pant RP. Common fixed points of the sequence of mappings. Ganita. 1996;**47**: 43-49

[16] Assad NA, Kirk WA. Fixed point theorems for set-valued mappings of contractive type. Pacific Journal of Mathematics. 1972;**43**:553-562

[17] Naimpally SA, Singh SL, Whitfield JHM. Fixed points in convex metric spaces. Mathematica Japonica. 1984;**29**:585-597

[18] Gähler S. 2-metric Raume und ihre topologiche Struktur. Mathematische Nachrichten. 1963;**26**:115-148

[19] Gähler S. Linear 2-normierte Räume. Mathematische Nachrichten. 1964;**28**:1-43

[20] Gähler S. Über 2-Banach Raume. Mathematische Nachrichten. 1969;**42**: 335-347

[21] Pant RP. R-weak commutativity and common fixed points of non-compatible maps. Ganita. 1998;**49**:19-27

[22] Nadler SB. Multivalued contraction mappings. Pacific Journal of Mathematics. 1969;**20**(2):457-488

[23] Ciric LJB. Fixed point for generalised multivalued contractions. Matematichki Vesnik. 1972;**9**(24):265-272

[24] Menger K. Statistical Metric. Proceedings of the National Academy Science of USA. 1942;**28**:535-537

[25] Sehgal VM. A fixed point theorem for mapping with a contractive iterate. Proceedings of the American Mathematical Society. 1969;**23**:631-634

[26] Sehgal VM, Bharucha-Reid AT. Fixed points of contraction mappings of probabilistic metric spaces. Mathematical Systems Theory. 1972;**6**:92-102

[27] Zadeh LA. Fuzzy Sets. Information and Control. 1965;**8**:338-353

[28] Zimmermann HJ, Sebastian HJ. Fuzzy design-Integration of fuzzy theory and knowledge-based system design. In: Proceedings of 1994 IEEE 3rd International Fuzzy Systems Conference. IEEE. 26-29 June 1994. DOI: 10.1109/FUZZY.1994.343673

[29] Zimmermann HJ, Sebastian HJ. Intelligent system design support by fuzzy multi-criteria decision making and evolutionary algorithms. In: Proceedings of the Fourth IEEE International Conference on Fuzzy Systems (FUZZ-IEEE/IFES 95, 20-24 March 1995). Yokohama, Japan: IEEE; 1995

[30] Tsourveloudis NC, Yannis A, Phillis A. Fuzzy assessment of machine flexibility. IEEE Transactions of Engineering Management. 1998;**45**(1):78-87

[31] Bedard R. Fixed point theorems for fuzzy number. Fuzzy Sets and Systems. 1984;**13**:291-302

[32] Butnariu D. Fixed points for fuzzy mappings. Fuzzy Sets and Systems. 1982; **7**:191-207

[33] Grabiec M. Fixed point in fuzzy metric space. Fuzzy Sets and Systems. 1988;**27**:385-389

[34] Weiss MD. Fixed points and induced fuzzy topologies for fuzzy sets. Journal of Mathematical Analysis and Applications. 1975;**50**:142-150

[35] Chang SSL, Zadeh L. On fuzzy mapping and control. IEEE Transactions on System Cybernetics. 1972;**2**:30-34

[36] Kandel A. Fuzzy dynamical systems and the nature of their solutions. In: Wang PP, Chang SK, editors. Fuzzy Sets Theory and Application to Policy Analysis and Information Systems. New York: Plenum Press; 1980. pp. 93-122

[37] Buckley JJ, Feuring T. Fuzzy differential equations. Fuzzy Sets and Systems. 2000;**110**:43-54

[38] Chalco-Cano Y, Roman-Flores H. Comparison between ´ some approaches to solve fuzzy differential equations. Fuzzy Sets and Systems. 2009;**160**: 1517-1527

[39] Nieto JJ, Rodriguez Lopez R. Euler polygonal method for ´ metric dynamical systems. Information Sciences. 2007;**177**: 587-600

[40] Prakash P, Sudha Priya G, Kim JH. Third-order three-point fuzzy boundary value problems. Nonlinear Analysis: Hybrid Systems. 2009;**3**:323-333

[41] Song S, Wu C. Existence and uniqueness of solutions to the Cauchy problem of fuzzy differential equations. Fuzzy Sets and Systems. 2000;**110**:55-67

[42] Kaleva O. Fuzzy differential equations. Fuzzy Sets and Systems. 1987; **24**:301-317

[43] Bede B, Gal SG. Almost periodic fuzzy-number-valued functions. Fuzzy Sets and Systems. 2004;**147**:385-403

[44] Bede B, Gal SG. Generalisations of the differentiability of fuzzy number value functions with applications to fuzzy differential equations. Fuzzy Sets and Systems. 2005;**151**:581-599

[45] Bede B, Rudas IJ, Bencsik AL. First order linear fuzzy differential equations under generalised differentiability. Information Sciences. 2007;**177**: 1648-1662

[46] Chalco Cano Y, Roman-Flores H. On new solutions of ′ fuzzy differential equations. Chaos, Solitons and Fractals. 2008;**38**:112-119

[47] Nieto JJ, Khastan A, Ivaz K. Numerical solution of fuzzy differential equations under generalised differentiability. Nonlinear Analysis: Hybrid Systems. 2009;**3**:700-707

[48] Allahviranloo T, Ahmady E, Ahmady N. Nth order fuzzy linear differential equations. Information Sciences. 2008;**178**:1309-1324

[49] Allahviranloo T, Abbasbandy S, Salahshour S, Hakimzadeh A. A new method for solving fuzzy linear differential equations. Computing. 2011; **92**:181-197

[50] Allahviranloo T, Ahmadi MB. Fuzzy Laplace transforms. Soft Computing. 2010;**14**:235-243

[51] Salahshour S, Allahviranloo T. Applications of fuzzy Laplace transform. Soft Computing. 2013;**17**:145-158. DOI: 10.1007/s00500-012-0907-4

Chapter 7

# Study of a Dynamical Problem under Fuzzy Conformable Differential Equation

*Atimad Harir, Said Melliani and Lalla Saadia Chadli*


## Abstract

The notion of inclusion by generalized conformable differentiability is used to analyze fuzzy conformable differential equations (FCDE). This idea is based on expanding the class of conformable differentiable fuzzy mappings, and we use generalized lateral conformable derivatives to do so. We'll see that both conformable derivatives are distinct and that they lead to different FCDE solutions. The approach's utility and efficiency are demonstrated with an example.


**Keywords:** fuzzy fractional differential equation, conformable fractional derivative, fuzzy number

## 1. Introduction

Aubin and Cellina [1] established the notion of differential inclusions systemically. They looked at the existence and qualities of differential inclusion solutions of the form [2].

$$u'(t) \in \Phi(u(t)) \quad or \quad u'(t) \in \Phi(t, u(t)). \tag{1}$$

In this paper we will consider the conformable fractional differential equation.

$$\begin{aligned} u^{(q)}(t) &= \Phi(t, u(t)) \\ u^{\kappa}(0) &\in [u_0]^{\kappa}, \quad \kappa \in [0,1] \end{aligned} \tag{2}$$

where $t \in (0, a)$ and $u_0$ is a fuzzy number. $u^{(q)}$ is the conformable fractional derivative of $u$ of order $\gamma \in (0,1]$ [3–5]. There are numerous options for defining a fuzzy fractional derivatives and, as a result, see [6–9], studying Eq. (2). [10–16] constructed the generalized derivative of a set value function and investigated it, while [17–20] explored the generalized conformable fractional derivative.

The objective of this research is to see if fuzzy solutions exist using conformable differential inclusion, using the generalized conformable differentiability concept

This idea is based on expanding the class of differentiable fuzzy mappings, and we use lateral conformable derivatives to do so. We will see that both derivatives are different and they lead us to different solutions from an FCDE.

## 2. Preliminaries

We'll go through a few definitions now that will come in handy later in the paper. Let us start with a definition. $\mathrm{R}_{\mathcal{F}}$ the class of fuzzy subsets of the real axis $\{\eta : \mathrm{R} \to [0,\ 1]\}$ satisfying the following properties:

i. $\eta$ is normal,

ii. $\eta$ is convex fuzzy set,

iii. $\eta$ is upper semicontinuous,

iv. $[\eta]^0 = cl\{x \in \mathrm{R} | \eta(x) > 0\}$ is compact.

Then $\mathrm{R}_{\mathcal{F}}$ is called the space of fuzzy numbers [21].

If $\eta$ is a fuzzy set, we define $[\eta]^\kappa = \{x \in \mathrm{R} | \eta(x) \geq \kappa\}$ the $\kappa$-level sets of $\eta$, with $0 < \kappa \leq 1$. Also, if $\eta \in \mathrm{R}_{\mathcal{F}}$ then $\kappa$-cut of $\eta$ denoted by $[\eta]^\kappa = [\eta_1^\kappa, \eta_2^\kappa]$.

For $\eta, \nu \in \mathrm{R}_{\mathcal{F}}$ and $\lambda \in \mathrm{R}$ the sum $\eta + \nu$ and the product $\lambda\eta$ are defined by $\forall \kappa \in [0, 1]$,

$$[\eta + \nu]^\kappa = [\eta_1^\kappa + \nu_1^\kappa, \eta_2^\kappa + \nu_2^\kappa], \tag{3}$$

$$[\lambda\eta]^\kappa = \lambda[\eta]^\kappa = \begin{cases} [\lambda\eta_1^\kappa, \lambda\eta_2^\kappa], & \lambda \geq 0; \\ [\lambda\eta_2^\kappa, \lambda\eta_1^\kappa], & \lambda < 0, \end{cases} \tag{4}$$

Let $d : \mathrm{R}_{\mathcal{F}} \times \mathrm{R}_{\mathcal{F}} \to \mathrm{R}_+ \cup \{0\}$ by the following equation [22].

$$d(\eta, \nu) = \sup_{\kappa \in [0,1]} d_H([\eta]^\kappa,\ [\nu]^\kappa),\ for\ all\ \eta, \nu \in \mathrm{R}_{\mathcal{F}}, \tag{5}$$

$$= \sup_{\kappa \in [0,1]} \max\{|\eta_1^\kappa - \nu_1^\kappa|,\ |\eta_2^\kappa - \nu_2^\kappa|\} \tag{6}$$

where $d_H$ is the Hausdorff metric.

The following properties are well-known see [22, 23]: $\forall\ \eta, \nu, \omega, \rho \in \mathrm{R}_{\mathcal{F}}$ and $\lambda \in \mathrm{R}$.

$$\begin{aligned} d(\eta + \omega, \nu + \omega) &= d(\eta, \nu) \quad and \quad d(\eta, \nu) = d(\nu, \eta), \\ d(\lambda\eta, \lambda\nu) &= |\lambda| d(\eta, \nu), \\ d(\eta + \nu, \omega + \rho) &\leq d(\eta, \omega) + d(\nu, \rho). \end{aligned} \tag{7}$$

And $(\mathrm{R}_{\mathcal{F}}, d)$ is a complete metric space.

**Theorem 1.** [1, 22] *Let $\eta : [0, a] \to \mathrm{R}_{\mathcal{F}}$ and $[\eta(t)]^\kappa = [\eta_1^\kappa(t), \eta_2^\kappa(t)]$ be Seikkala differentiable. Then, $\eta_1^\kappa(t)$ and $\eta_2^\kappa(t)$ are differentiable and*

$$[\eta'(t)]^\kappa = \left[(\eta_1^\kappa)'(t),\ (\eta_2^\kappa)'(t)\right],\ \kappa \in [0, 1]. \tag{8}$$

**Definition 1.** [24] *Let* $\eta : [0, a] \to \mathrm{R}_{\mathcal{F}}$. $\int_b^c \eta(t)dt, \quad b, c \in [0, a]$ *is the fuzzy integral, defined by*

$$\left[\int_b^c \eta(t)dt\right]^{\kappa} = \left[\int_b^c \eta_1^{\kappa}(t)dt, \ \int_b^c \eta_2^{\kappa}(t)dt\right], \tag{9}$$

for all $0 \le \kappa \le 1$. In [24], if $\eta : [0, a] \to \mathrm{R}_{\mathcal{F}}$ is continuous, it is fuzzy integrable.
**Theorem 2.** [22, 25] *If* $\eta \in \mathrm{R}_{\mathcal{F}}$, *then:*

i.

$$[\eta]^{\kappa_2} \subset [\eta]^{\kappa_1}, \quad \textit{if} \ \ 0 \le \kappa_1 \le \kappa_2 \le 1; \tag{10}$$

ii. $\{\kappa_k\} \subset [0, 1]$ *is a increasing sequence which converges to* $\kappa$,

$$[\eta]^{\kappa} = \bigcap_{k \ge 1} [\eta]^{\kappa_k}. \tag{11}$$

Alternatively, if $\Upsilon_\kappa = \left\{ \left[\eta_1^\kappa, \eta_2^\kappa\right]; \ \kappa \in (0, 1] \right\}$ is a closed real intervals (i) and (ii), then $\{\Upsilon_\kappa\}$ defined a fuzzy number $\eta \in \mathrm{R}_{\mathcal{F}}$ such that $[\eta]^{\kappa} = \Upsilon_\kappa$..

## 3. Fuzzy conformable differentiability and integral

The funcion $\Phi : [a, b] \to \mathbb{R}_{\mathcal{F}}$ is called fuzzy function. The $\kappa$-level representation of fuzzy function $\Phi$ given by $[\Phi(t)]^{\kappa} = \left[\phi_1^{\kappa}(t), \phi_2^{\kappa}(t)\right], \ \forall t \in [a, b], \ \forall \kappa \in [0, 1]$.

**Definition 2.** [17] *Let* $\Phi : (0, a) \to \mathbb{R}_{\mathcal{F}}$ *be a fuzzy function.* $\gamma^{th}$ *order" fuzzy conformable derivative" of* $\Phi$ *is defined by*

$$T_\gamma(\Phi)(t) = \lim_{\varepsilon \to 0^+} \frac{\Phi(t + \varepsilon t^{1-\gamma}) \ominus \Phi(t)}{\varepsilon} = \lim_{\varepsilon \to 0^+} \frac{\Phi(t) \ominus \Phi(t - \varepsilon t^{1-\gamma})}{\varepsilon}. \tag{12}$$

for all $t > 0, \gamma \in (0, 1)$. Let $\Phi^{(\gamma)}(t)$ stands for $T_\gamma(\Phi)(t)$. Hence

$$\Phi^{(\gamma)}(t) = \lim_{\varepsilon \to 0^+} \frac{\Phi(t + \varepsilon t^{1-\gamma}) \ominus \Phi(t)}{\varepsilon} = \lim_{\varepsilon \to 0^+} \frac{\Phi(t) \ominus \Phi(t - \varepsilon t^{1-\gamma})}{\varepsilon}. \tag{13}$$

If $\Phi$ is $\gamma$-differentiable in some $(0, \ a)$, and $\lim_{t \to 0^+} \Phi^{(\gamma)}(t)$ exists, then

$$\Phi^{(\gamma)}(0) = \lim_{t \to 0^+} \Phi^{(\gamma)}(t) \tag{14}$$

and the limits (in the metric d).
**Remark 1.** [26]

— *If* $\Phi$ *is* $\gamma$*-differentiable then the multivalued mapping* $\Phi_\kappa$ *is* $\gamma$*-differentiable for all* $\kappa \in [0, 1]$ *and*

$$T_\gamma \Phi_\kappa = \left[\Phi^{(\gamma)}(t)\right]^{\kappa}, \tag{15}$$

where $T_\gamma \Phi_\kappa$ is denoted from the conformable fractional derivative of $\Phi_\kappa$ of order $\gamma$.

— $[\eta]^\kappa \ominus [\nu]^\kappa$, $\kappa \in [0,1]$ *does not imply the existence of Hukuhara difference (H-difference)* $\eta \ominus \nu$.

**Theorem 3.** [26]
Let $\Phi : (0,a) \rightarrow \mathbb{R}_{\mathcal{F}}$, $[\Phi(t)]^\kappa = [\phi_1^\kappa(t), \phi_2^\kappa(t)]$, $\kappa \in [0,1]$.

i. *If* $\Phi$ *is* $\gamma_{(1)}$*-differentiable, then* $\phi_1^\kappa(t)$ *and* $\phi_2^\kappa(t)$ *are* $\gamma$*-differentiable and*

$$\left[\Phi^{(\gamma_{(1)})}(t)\right]^\kappa = \left[\left(\phi_1^\kappa\right)^{(\gamma)}(t), \left(\phi_2^\kappa\right)^{(\gamma)}(t)\right] \tag{16}$$

ii. *If* $\Phi$ *is* $\gamma_{(2)}$*-differentiable, then* $\phi_1^\kappa(t)$ *and* $\phi_2^\kappa(t)$ *are* $\gamma$*-differentiable and*

$$\left[\Phi^{(\gamma_{(2)})}(t)\right]^\kappa = \left[\left(\phi_2^\kappa\right)^{(\gamma)}(t), \left(\phi_1^\kappa\right)^{(\gamma)}(t)\right]. \tag{17}$$

**Theorem 4.** [17] *Let* $\gamma \in (0,1]$ :.

i. *If* $\Phi$ *is* (1)*-differentiable and* $\Phi$ *is* $\gamma_{(1)}$*-differentiable, then*

$$T_{\gamma_{(1)}}\Phi(t) = t^{1-\gamma} D_1^1 \Phi(t) \tag{18}$$

ii. *If* $\Phi$ *is* (2)*-differentiable and* $\Phi$ *is* $\gamma_{(2)}$*-differentiable, then*

$$T_{\gamma_{(2)}}\Phi(t) = t^{1-\gamma} D_2^1 \Phi(t) \tag{19}$$

**Theorem 5.** *If* $\Phi : (0,a) \rightarrow \mathbb{R}_{\mathcal{F}}$ *is* $\gamma$*-differentiable then it is continuous.*
*Proof.* Denote $\Phi_\kappa(t) = [\phi_1^\kappa(t), \phi_2^\kappa(t)], \kappa \in [0,1]$. Then $\phi_1^\kappa(t)$ and $\phi_2^\kappa(t)$ are continuous at $t_0$, so $\Phi$ is continuous at $t_0$.
If $\varepsilon > 0$ and $\kappa \in [0,1]$, we have:

$$\left[\Phi\left(t_0 + \varepsilon t_0^{1-\gamma}\right) \ominus \Phi(t_0)\right]^\kappa = \left[\phi_1^\kappa\left(t_0 + \varepsilon t_0^{1-\gamma}\right) - \phi_1^\kappa(t_0), \phi_2^\kappa\left(t_0 + \varepsilon t_0^{1-\gamma}\right) - \phi_2^\kappa(t_0)\right]$$

Dividing and multiplying by $\varepsilon$, we have:

$$\left[\Phi\left(t_0 + \varepsilon t_0^{1-\gamma}\right) \ominus \Phi(t_0)\right]^\kappa = \left[\frac{\phi_1^\kappa\left(t_0 + \varepsilon t_0^{1-\gamma}\right) - \phi_1^\kappa(t_0)}{\varepsilon} \cdot \varepsilon, \frac{\phi_2^\kappa\left(t_0 + \varepsilon t_0^{1-\gamma}\right) - \phi_2^\kappa(t_0)}{\varepsilon} \cdot \varepsilon\right]$$

Similarly, we obtain:

$$\left[\Phi(t_0) \ominus \Phi\left(t_0 - \varepsilon t_0^{1-\gamma}\right)\right]^\kappa = \left[\frac{\phi_1^\kappa(t_0) - \phi_1^\kappa\left(t_0 - \varepsilon t_0^{1-\gamma}\right)}{\varepsilon} \cdot \varepsilon, \frac{\phi_2^\kappa(t_0) - \phi_2^\kappa\left(t_0 - \varepsilon t_0^{1-\gamma}\right)}{\varepsilon} \cdot \varepsilon\right]$$

Then

$$\lim_{\varepsilon\to 0^+}\left[\Phi\left(t_0+\varepsilon t_0^{1-\gamma}\right)\ominus\Phi(t_0)\right]^\kappa = \left[\lim_{\varepsilon\to 0^+}\frac{\phi_1^\kappa\left(t_0+\varepsilon t_0^{1-\gamma}\right)-\phi_1^\kappa(t_0)}{\varepsilon}\cdot\lim_{\varepsilon\to 0^+}\varepsilon,\right.$$
$$\left.\lim_{\varepsilon\to 0^+}\frac{\phi_2^\kappa\left(t_0+\varepsilon t_0^{1-\gamma}\right)-\phi_2^\kappa(t_0)}{\varepsilon}\cdot\lim_{\varepsilon\to 0^+}\varepsilon\right]$$

Similarly, we obtain:

$$\lim_{\varepsilon\to 0^+}\left[\Phi(t_0)\ominus\Phi\left(t_0-\varepsilon t_0^{1-\gamma}\right)\right]^\kappa = \left[\lim_{\varepsilon\to 0^+}\frac{\phi_1^\kappa(t_0)-\phi_1^\kappa\left(t_0-\varepsilon t_0^{1-\gamma}\right)}{\varepsilon}\cdot\lim_{\varepsilon\to 0^+}\varepsilon,\right.$$
$$\left.\lim_{\varepsilon\to 0^+}\frac{\phi_2^\kappa(t_0)-\phi_2^\kappa\left(t_0-\varepsilon t_0^{1-\gamma}\right)}{\varepsilon}\cdot\lim_{\varepsilon\to 0^+}\varepsilon\right]$$

Let $h=\varepsilon t_0^{1-\gamma}$. Then

$$\lim_{h\to 0^+}[\Phi(t_0+h)\ominus\Phi(t_0)]^\kappa=\left[\left(\phi_1^\kappa\right)^{(\gamma)}(t_0)\cdot 0,\left(\phi_2^\kappa\right)^{(\gamma)}(t_0)\cdot 0\right]$$

Similarly, we obtain:

$$\lim_{h\to 0^+}[\Phi(t_0-h)]^\kappa=[\Phi(t_0)]^\kappa$$

which implies that

$$\lim_{h\to 0^+}[\Phi(t_0+h)]^\kappa=[\Phi(t_0)]^\kappa$$

Similary, we obtain:

$$\lim_{h\to 0^+}[\Phi(t_0-h)]^\kappa=[\Phi(t_0)]^\kappa$$

Hence, $\Phi$ is continuous at $t_0$. □

**Remark 2.** *If* $\Phi:(0,a)\to\mathbb{R}_\mathcal{F}$ *is* $\gamma$*-differentiable and* $\Phi^{(\gamma)}$ *for all* $\gamma\in(0,1]$ *is continuous, then we denote* $\Phi\in C^1((0,a),\mathbb{R}_\mathcal{F})$.

**Theorem 6.** *Let* $\gamma\in(0,1]$ *and if* $\Phi,\Psi:(0,a)\to\mathbb{R}_\mathcal{F}$ *are* $\gamma$*-differentiable and* $\lambda\in\mathbb{R}$ *then*

i.
$$T_\gamma(\Phi+\Psi)(t)=T_\gamma(\Phi)(t)+T_\gamma(\Psi)(t) \tag{20}$$

ii.
$$T_\gamma(\lambda\Phi)(t)=\lambda T_\gamma(\Phi)(t). \tag{21}$$

*proof.* We present the details only for case (i), since the other case is anlogous. Since $\Phi$ is $\gamma_{(1)}$-differentiable it follows that $\Phi(t+\varepsilon t^{1-\gamma})\ominus\Phi(t)$ exists i.e. there exists $u_1(t,\varepsilon t^{1-\gamma})$ such that

$$\Phi\left(t+\varepsilon t^{1-\gamma}\right)=\Phi(t)+u_1\left(t,\varepsilon t^{1-\gamma}\right) \tag{22}$$

Analogously since $\Psi$ is $\gamma_{(1)}$-differentiable there exists $v_1(t,\varepsilon t^{1-\gamma})$ such that

$$\Psi\left(t+\varepsilon t^{1-\gamma}\right)=\Psi(t)+v_1\left(t,\varepsilon t^{1-\gamma}\right)$$

and we get

$$\Phi\left(t+\varepsilon t^{1-\gamma}\right)+\Psi\left(t+\varepsilon t^{1-\gamma}\right)=\Phi(t)+\Psi(t)+u_1\left(t,\varepsilon t^{1-\gamma}\right)+v_1\left(t,\varepsilon t^{1-\gamma}\right) \tag{23}$$

that is the H-difference

$$\left(\Phi\left(t+\varepsilon t^{1-\gamma}\right)+\Psi\left(t+\varepsilon t^{1-\gamma}\right)\right)\ominus\left(\Phi(t)+\Psi(t)\right)=u_1\left(t,\varepsilon t^{1-\gamma}\right)+v_1\left(t,\varepsilon t^{1-\gamma}\right) \tag{24}$$

By similar reasoning we get that there exist $u_2(t,\varepsilon t^{1-\gamma})$ and $v_2(t,\varepsilon t^{1-\gamma})$ such that

$$\begin{aligned}\Phi(t)&=\Phi\left(t-\varepsilon t^{1-\gamma}\right)+u_2\left(t,\varepsilon t^{1-\gamma}\right)\\ \Psi(t)&=\Psi\left(t-\varepsilon t^{1-\gamma}\right)+v_2\left(t,\varepsilon t^{1-\gamma}\right)\end{aligned}$$

and so

$$\left(\Phi(t)+\Psi(t)\right)=\left(\Phi\left(t-\varepsilon t^{1-\gamma}\right)+\Psi\left(t-\varepsilon t^{1-\gamma}\right)\right)+u_2\left(t,\varepsilon t^{1-\gamma}\right)+v_2\left(t,\varepsilon t^{1-\gamma}\right)$$

that is the H-difference

$$\left(\Phi(t)+\Psi(t)\right)\ominus\left(\Phi\left(t-\varepsilon t^{1-\gamma}\right)+\Psi\left(t-\varepsilon t^{1-\gamma}\right)\right)=u_2\left(t,\varepsilon t^{1-\gamma}\right)+v_2\left(t,\varepsilon t^{1-\gamma}\right) \tag{25}$$

We observe that

$$\begin{aligned}\lim_{\varepsilon\to 0^+}\frac{u_1\left(t,\varepsilon t^{1-\gamma}\right)}{\varepsilon}&=\lim_{\varepsilon\to 0^+}\frac{u_2\left(t,\varepsilon t^{1-\gamma}\right)}{\varepsilon}=\Phi^{(\gamma)}(t)\quad\text{and}\\ \lim_{\varepsilon\to 0^+}\frac{v_1\left(t,\varepsilon t^{1-\gamma}\right)}{\varepsilon}&=\lim_{\varepsilon\to 0^+}\frac{v_2\left(t,\varepsilon t^{1-\gamma}\right)}{\varepsilon}=\Psi^{(\gamma)}(t).\end{aligned}$$

Finally, by multiplying (24) and (25) with $\frac{1}{\varepsilon}$ and passing to limit with $\lim_{\varepsilon\to 0^+}$ we get that $\Phi+\Psi$ is $\gamma_{(1)}$-differentiable and $T_\gamma(\Phi+\Psi)(t)=T_\gamma\Phi(t)+T_\gamma\Psi(t)$ The case when $\Phi$ and $\Psi$ are $\gamma_{(2)}$-differentiable is similar to the previous one. □

**Definition 3.** *Let $\Phi\in C((0,\ a),\ \mathbb{R}_{\mathcal{F}})\cap L^1((0,\ a),\ \mathbb{R}_{\mathcal{F}})$, Define the fuzzy fractional integral for $\gamma\in(0,1]$.*

$$I_\gamma(\Phi)(t)=I_1\left(t^{\gamma-1}\Phi\right)(t)=\int_0^t\frac{\Phi}{s^{1-\gamma}}(s)ds, \tag{26}$$

where the integral is the usual Riemann improper integral.

**Theorem 7.** *$T_\gamma I_\gamma(\Phi)(t)$, for $t\geq 0$, where $\Phi$ is any continuous function in the domain of $I_\gamma$.*

*Proof.* Since $\Phi$ is continuous, then $I_\gamma(\Phi)(t)$ is clearly conformable differentiable. Hence,

*DOI: http://dx.doi.org/10.5772/intechopen.105904*

$$
\begin{aligned}
[T_\gamma I_\gamma(\Phi)(t)]^\kappa &= \left[t^{1-\gamma}\frac{d}{dt}I_\gamma(\Phi)(t)\right]^\kappa \\
&= \left[t^{1-\gamma}\frac{d}{dt}\int_0^t \frac{\phi_1^\kappa(x)}{x^{1-\gamma}}dx, t^{1-\gamma}\frac{d}{dt}\int_0^t \frac{\phi_2^\kappa(x)}{x^{1-\gamma}}dx\right] \\
&= \left[t^{1-\gamma}\frac{\phi_1^\kappa(t)}{t^{1-\gamma}}, t^{1-\gamma}\frac{\phi_2^\kappa(t)}{t^{1-\gamma}}\right] \\
&= [\Phi(t)]^\kappa
\end{aligned}
$$

□

**Theorem 8.** *Let $\gamma \in (0,1]$ and $\Phi$ be $\gamma$-differentiable in $(0,a)$ and assume that the conformable derivative $\Phi^{(\gamma)}$ is integrable over $(0,a)$. Then $\forall\ s \in (0,a)$ we have.*

$$\Phi(s) = \Phi(a) + I_\gamma \Phi^{(\gamma)}(t) \tag{27}$$

*Proof.* Let $\gamma \in (0,1]$ and $\kappa \in [0,1]$ be fixed. We will demonstrate this.

$$\Phi_\kappa(s) = \Phi_\kappa(a) + I_\gamma \Phi_\kappa^{(\gamma)} \tag{28}$$

where $\Phi_\kappa^{(\gamma)}$ is conformable derivative of $\Phi_\kappa$, the equation is then obtained by applying Theorems 3 and 4.

$$
\begin{aligned}
\Phi_\kappa(s) &= \Phi_\kappa(a) + I_\gamma \Phi_\kappa^{(\gamma)} \\
&= \Phi_\kappa(a) + I_\gamma\left(t^{1-\gamma}\Phi_\kappa'\right)
\end{aligned}
$$

by (26) we have

$$
\begin{aligned}
\Phi_\kappa(s) &= \Phi_\kappa(a) + I_\gamma\left(t^{1-\gamma}\Phi_\kappa'\right) \\
&= \Phi_\kappa(a) + \int_0^s t^{\gamma-1}\left(t^{1-\gamma}\Phi_\kappa'\right)
\end{aligned}
$$

So

$$\Phi_\kappa(s) = \Phi_\kappa(a) + \int_0^s \Phi_\kappa' \tag{29}$$

where $\Phi_\kappa'$ is the derivative of $\Phi_\kappa$, For a fuzzy mapping, the (29) is likewise true $\Phi : (0,a) \to \mathbb{R}_\mathcal{F}$. In [1], the equality (28) now follows Theorem (8).

## 4. Solutions via conformable differential inclusions

We consider the fuzzy conformable differential equation.

$$u^{(\gamma)}(t) = \Phi(t, u(t)), \quad u(0) = u_0, \ \ \gamma \in (0,1], \tag{30}$$

where $\Phi : (0,a) \times \mathrm{R}_\mathcal{F} \to \mathrm{R}_\mathcal{F}$ is generated from a continuous function using Zadeh's extension principle. $\psi : (0,a) \times \mathrm{R} \to \mathrm{R}$.

Let $\Phi(t,u)$ can be calculated at the level, i.e., $\forall\ \kappa \in [0,1]$.

$$[\Phi(t,u)]^{\kappa} = \psi(t,[u]^{\kappa})$$

for all $t \in (0,a)$, $u \in \mathrm{R}_{\mathcal{F}}$ and $\kappa \in [0,1]$. We interpret the fuzzy initial value problem (30) as a set of differential inclusions, following Diamonds [7, 10].

$$\left(v^{(\gamma)}\right)^{\kappa}(t) = \psi(t, v^{\kappa}(t)), \quad v^{\kappa}(0) \in [u_0]^{\kappa} \tag{31}$$

The reachable sets, under reasonable assumptions.

$$\Upsilon_{\kappa}(t) = \{v^{\kappa}(t) |\ v^{\kappa} \ \textit{is a solution of} \ (31)\},$$

are $\kappa$-cuts of a fuzzy set $u(t)$, which we call a solution of (30). If we assume that the solutions to the initial value problem are unique, $\left(v^{(\gamma)}\right)^{\kappa}(t) = \psi(t, v^{\kappa}(t))$, $\quad v^{\kappa}(0) = v_0$, it follows that $\Upsilon_{\kappa}(t) = [w_1(t), w_2(t)]$, where.

$$\left(w^{(\gamma)}\right)_1(t) = \psi(t, w_1(t)), \quad w_1(0) = u_{01}^{\kappa} \quad \text{and}$$
$$\left(w^{(\gamma)}\right)_2(t) = \psi(t, w_2(t)), \quad w_2(0) = u_{02}^{\kappa}$$

**Theorem 9.** *The fuzzy solution and solution by differential inclusions solution using the first form are equivalent if $\psi$ is nondecreasing with respect to the second argument.*

*proof.* For each $\kappa \in [0,1]$ and $\forall \ \gamma \in (0,1]$, we think $[u(t))]^{\kappa} = \left[u_1^{\kappa}(t), u_2^{\kappa}(t)\right]$ and $[u(0))]^{\kappa} = \left[u_{01}^{\kappa}, u_{02}^{\kappa}\right]$. Since $g$ is continuous and

$$[\Phi(t,u(t))]^{\kappa} = \psi(t,[u(t))]^{\kappa}) = \psi\left(t, \left[u_1^{\kappa}(t), u_2^{\kappa}(t)\right]\right),$$

then $\psi\left(t, \left[u_1^{\kappa}(t), u_2^{\kappa}(t)\right)\right]$ is compact and connected i.e. a closed bounded interval. As $\psi$ is nondecreasing, we have that

$$\psi\left(t, \left[u_1^{\kappa}(t), u_2^{\kappa}(t)\right]\right) = \left[\psi\left(t, u_1^{\kappa}(t)\right), \psi\left(t, u_2^{\kappa}(t)\right)\right]$$

As a result, the conformable differential system for boundary functions of fuzzy solution is uncoupled into two initial value problems:

$$\left(u^{(\gamma)}\right)_1^{\kappa}(t) = \psi\left(t, u_1^{\kappa}(t)\right), \quad u_1^{\kappa}(0) = u_{01}^{\kappa},$$
$$\left(u^{(\gamma)}\right)_2^{\kappa}(t) = \psi\left(t, u_2^{\kappa}(t)\right), \quad u_2^{\kappa}(0) = u_{02}^{\kappa},$$

This results in $u_1^{\kappa} = w_1$ and $u_2^{\kappa} = w_2$.

Now, we offer the following result as an extension of the preceding theorem to the class of differentiable functions with regard to the second form (17).

**Theorem 10.** *If $\psi$ is nonincreasing with respect to the second argument then, using the derivative in the second form* (17), *the fuzzy solution of* (30) *and the solution via differential inclusions are identical.*

*proof.* Let $\kappa \in [0,1]$ and $\gamma \in (0,1]$, we consider.

$$[u(t))]^{\kappa} = \left[u_1^{\kappa}(t), u_2^{\kappa}(t)\right] \quad \textit{and} \quad [u(0))]^{\kappa} = \left[u_{01}^{\kappa}, u_{02}^{\kappa}\right].$$

So

$$[\Phi(t,u(t))]^{\kappa} = \psi(t,[u(t))]^{\kappa}) = \psi\left(t,\left[u_1^{\kappa}(t),u_2^{\kappa}(t)\right]\right)$$

and $\psi$ is continuous, and $\psi\left(t,\left[u_1^{\kappa}(t),u_2^{\kappa}(t)\right]\right)$ is a closed bounded interval. Since $\psi$ is nonincreasing, it follows that

$$\psi\left(t,\left[u_1^{\kappa}(t),u_2^{\kappa}(t)\right]\right) = \left[\psi\left(t,u_2^{\kappa}(t)\right),\psi\left(t,u_1^{\kappa}(t)\right)\right]$$

Consequently, from (31), we have the conformable differential system.

$$\begin{aligned}\left(u^{(\gamma)}\right)_2^{\kappa}(t) &= \psi\left(t,u_2^{\kappa}(t)\right), \quad u_2^{\kappa}(0)=u_{02}^{\kappa},\\ \left(u^{(\gamma)}\right)_1^{\kappa}(t) &= \psi\left(t,u_1^{\kappa}(t)\right), \quad u_1^{\kappa}(0)=u_{01}^{\kappa},\end{aligned} \tag{32}$$

If $u^{(\gamma)}(t)$ is consider in the form (2), we have that

$$\left[\left(u^{(\gamma)}\right)(t)\right]^{\kappa} = \left[\left(u^{(\gamma)}\right)_2^{\kappa}(t),\left(u^{(\gamma)}\right)_1^{\kappa}(t)\right] = \left[\psi\left(t,u_2^{\kappa}(t)\right),\psi\left(t,u_1^{\kappa}(t)\right)\right]$$

The proof is complete after obtaining the differential system (32).
**Example 1.** *Consider the fuzzy initial value problem.*

$$u^{(\gamma)}(t) = u^2(t), \quad u(0)=u_0 \tag{33}$$

Where $u_0$ is a traingular fuzzy number $[u_0]^{\kappa} = [1+\kappa, 3-\kappa]$. Since $u^2$ is continuous and we are operating on $R_{\mathcal{F}}$, we can solve the equation levelwise.

Since $u^2$ is increasing when $x>0$, we have to solve a conformable differential system for $\gamma \in (0,1]$.

$$\left(u_1^{\kappa}\right)^{(\gamma)}(t) = \left(u_1^{\kappa}\right)^2(t), \quad u_1^{\kappa}(0)=1+\kappa, \tag{34}$$

$$\left(u_2^{\kappa}\right)^{(\gamma)}(t) = \left(u_2^{\kappa}\right)^2(t), \quad u_2^{\kappa}(0)=3-\kappa, \tag{35}$$

where $[u(t)]^{\kappa} = \left[u_1^{\kappa},u_2^{\kappa}\right]$.
The solutions are

$$u_1^{\kappa}(t) = \frac{-1-\kappa}{\left(\frac{t^{\gamma}}{\gamma}+\frac{t^{\gamma}}{\gamma}\kappa\right)-1} \text{ and } u_2^{\kappa}(t) = \frac{-3+\kappa}{\left(3\frac{t^{\gamma}}{\gamma}-\frac{t^{\gamma}}{\gamma}\kappa\right)-1}$$

We can see this $u_2^{\kappa}(t)<\infty$ for $t<\frac{1}{3}$ and

$$0 \le u_1^{\kappa}(t) \le u_2^{\kappa}(t)$$

for these values of $t$.

As a result, there is a fuzzy solution to the fuzzy initial value problem. $u(t)$ for $0 \le t < \frac{1}{3}$.

## 5. Conclusion

The fuzzy conformable differential inclusions (FCDI) are introduced, which have been used by various authors to solve FDE for $\gamma = 1$ [2, 6]. It also has the advantage that the solutions derived using FCDI appear to be more intuitive than other conformable derivative solutions first form (9), [19]. It's also worth noting that this interpretation has a drawback in that we cannot discuss the fuzzy conformable derivative. Instead, we address this obstacle by utilizing the fuzzy conformable derivative in the second form (17), and the fuzzy solution and the solution via conformable differential inclusions are identical.

## Conflicts of interest

The authors declare that they have no conflicts of interest.

## Data availability

The data used to support the findings of this study are available from the corresponding author upon request.

## Author details

Atimad Harir*, Said Melliani and Lalla Saadia Chadli
Laboratory of Applied Mathematics and Scientific Computing, Sultan Moulay Slimane University, Beni Mellal, Morocco

*Address all correspondence to: atimad.harir@gmail.com

## References

[1] Kaleva O. A note on fuzzy differential equations. Nonlinear Analysis. 2006;**64**: 895-900

[2] Diamond P. Time-dependent differential inclusions, cocycle attractors and fuzzy differential equations. IEEE Transactions on Fuzzy Systems. 1999;7: 734-740

[3] Abdeljawad T. On conformable fractional calculus. Journal of Computational and Applied Mathematics. 2015;**279**:57-66

[4] Khalil R, Al Horani M, Yousef A, Sababheh M. A new definition of fractional derivative. Journal of Computational and Applied Mathematics. 2014;**264**:65-70

[5] Unal E, Gokdogan A. Solution of conformable fractional ordinary differential equations via differential transform method. International Journal for Light and Electron Optics. 2017;**128**: 264-273

[6] Abbasbandy S, Nieto JJ, Alavi M. Tuning of reachable set in one dimensional fuzzy differential inclusions. Chaos, Solitons-Fractals. 2005;**26**:1337-1341

[7] Arshad S, Lupulescu V. On the fractional differential equations with uncertainty. Nonliniear Analysis. 2011; **74**:3685-3693

[8] Harir A, Melliani S, Chadli LS, Minchev E. Solutions of fuzzy fractional heatlike and wave-like equations by variational iteration method. International Journal of Contemporary Mathematical Sciences. 2020;**15**(1):11-35

[9] Harir A, Melliani S, Chadli LS. Fuzzy fractional evolution equations and fuzzy solution operators. Advanced Fuzzy Systems. 2019;**2019**:10. DOI: 10.1155/2019/5734190

[10] Bede B, Gal SG. Almost periodic fuzzy number valued functions. Fuzzy Sets and Systems. 2004;**147**:385-403

[11] Bede B, Gal SG. Generalizations of the differentiability of fuzzy number value functions with applications to fuzzy differential equations. Fuzzy Sets and Systems. 2005;**151**:581-599

[12] Goo HY, Park JS. On the continuity of the Zadeh extensions. Journal of the Chungcheong Mathematical Society. 2007;**20**(4):525-533

[13] Shah K, Arfan M, Ullah A, Mdallal Q, Ansari KJ, Abdeljawad T. Computational study on the dynamics of fractional order differential equations with applications. Chaos, Solitons and Fractals. 2022;**157**:111955

[14] Shah K, Naz H, Sarwar M, Abdeljawad T. On spectral numerical method for variable-order partial differential equations. AIMS Mathematics. 2022;**7**(6):10422-10438

[15] Shah K, Ali A, Zeb S, Khan A, Alqudah MA, Abdeljawad T. Study of fractional order dynamics of nonlinear mathematical model. Alexandria Engineering Journal. 2022;**61**(12): 11211-11224

[16] Shahid A, Khan A, Shah K, Alqudah MA, Abdeljawad T, Islam Su. On computational analysis of highly nonlinear model addressing real world applications. Results in Physics. 2022;**36**:105431

[17] Harir A, Melliani S, Chadli LS. Fuzzy generalized conformable fractional derivative. Advanced Fuzzy Systems. 2019;**2020**:7. DOI: 10.1155/2020/1954975

[18] Seikkala S. On the fuzzy initialvalue problem. Fuzzy Sets and Systems. 1987; **24**:319-330

[19] Harir A, Melliani S, Chadli LS. The fractional differential equations with uncertainty by conformable derivative. European Journal of Pure and Applied Mathematics. 2022;**15**(2):557-571. DOI: 10.29020/nybg.ejpam.v15i2.4299

[20] Harir A, Melliani S, Chadli LS. Fuzzy conformable fractional semigroups of operators. International Journal of Differential Equations. 2020, 2020:6. DOI: 10.1155/2020/8836011

[21] Diamond P, Kloeden PE. Metric Spaces of Fuzzy Sets: Theory and Applications. Singapore: World Scienific; 1994

[22] Harir A, Melliani S, Chadli LS. Existence, uniqueness and approximate solutions of fuzzy fractional differential equations. In: Fuzzy Systems—Theory and Applications. London, UK: IntechOpen; 2020. DOI: 10.5772/intechopen.94000

[23] Puri ML, Ralescu DA. Differentials of fuzzy functions. Journal of Mathematical Analysis and Applications. 1983;**91**:552-558

[24] Song S, Guo L, Feng C. Global existence of solutions to fuzzy differential equations. Fuzzy Sets and Systems. 2000;**115**:371-376

[25] Negoita CV, Ralescu DA. Applications of Fuzzy Sets to System Analysis. Basel: Birkhauser; 1975

[26] Harir A, Melliani S, Chadli LS. Analytic solution method for fractional fuzzy conformable Laplace transforms. SeMA. 2021;**78**:401-414. DOI: 10.1007/s40324-021-00240-7

Chapter 8

# Electrical Circuits as Dynamical Systems

*Alexandru G. Gheorghe and Mihai E. Marin*

## Abstract

An electrical circuit containing at least one dynamic circuit element (inductor or capacitor) is an example of a dynamic system. The behavior of inductors and capacitors is described using differential equations in terms of voltages and currents. The resulting set of differential equations can be rewritten as state equations in normal form. The eigenvalues of the state matrix can be used to verify the stability of the circuit. The most fitted numerical methods to integrate electrical circuit differential equations are the Euler Method (Forward and Backward), the Trapezoidal Rule, and the Gear Method of second to sixth degree, for circuits having stiff equations. These methods are implemented, with adjustable time-step integration, in the majority of circuit simulation software, such as SPICE. The analytical solution can also be computed, for small-size circuits, applying the Laplace Transform. It is interesting to compare the graphical presentation of numerically and analytically obtained solutions. While the numerical methods can be used for both linear and nonlinear circuits, the Laplace Transform is mostly used for linear circuits. A method of using it for nonlinear circuits is also presented.

**Keywords:** electrical circuits, state equations, Laplace Transform, numerical methods, linear and nonlinear circuits

## 1. Introduction

The existence and uniqueness of a dynamic circuit solution are strongly tied to the existence of the state equation in normal form. If the equation $\dot{x} = f(x, t)$ exists, then it can be proved, as it can be found in the mathematic literature, that if $f$ is Lipschitzian (for any $x_1$ and $x_2$ and any $t$, $\|f(x_1, t) - f(x_2, t)\| \leq k \cdot \|x_1 - x_2\|$ where $k > 0$ and $\|\|$ is the Euclidian norm) and if the function $f(0, t)$ is uniformly continuous and bounded, then the state equation has an unique solution for any initial state $x_1 = x(t)$. The existence of the normal form of the state equation is related to the existence and uniqueness of the resistive multiport solution for any values of the source parameters (which replace the dynamic elements) connected to the ports, and the existence of nonzero dynamic capacitances and inductances for any values of the control parameters of these elements.

It can be seen as in the case of linear circuits $f(x, t)$ is Lipschitzian: $\|f(x_1, t) - f(x_2, t)\| = \|A \cdot (x_1 - x_2)\|$ because there is always a constant $k$ such that $\|A \cdot (x_1 - x_2)\| \leq k \cdot \|x_1 - x_2\|$ [1–3].

In the case of nonlinear circuits without excess state quantities (the circuit does not contain any loop consisting only of independent or controlled voltage sources and/or capacitors and does not contain any cut-set consisting only of independent or controlled current sources and/or inductors), if the characteristics of the dynamic elements are strictly increasing and derivable, then they have a nonzero dynamic parameter at any point of operation. If the resistors' characteristics are strictly increasing and the resistive multiport does not have loops formed only from voltage sources and cut-sets formed only from current sources, then this resistive multiport has a unique solution. So, there are state equations in the normal form $\dot{x} = f(x, t)$. For the dynamic circuit to have a unique solution for any initial state $x(t_0)$, it is enough that $f$ is Lipschitzian [1–3]. When we have excess state quantities, the problem is treated similarly.

## 2. State equations for dynamic circuits

The simplest *dynamic* circuit elements are the linear capacitor and the linear inductor. The operating equation of the linear capacitor is $i_c(t) = C \cdot \frac{dv_c(t)}{dt}$ where $v_c(t)$ is the voltage at the capacitor terminals, $i_c(t)$ is the current through the capacitor, and $C$ is a constant called the capacitor capacity. The operating equation of the linear inductor is $v_L(t) = L \cdot \frac{di_L(t)}{dt}$ where $v_L$ is the voltage at the inductor terminals, $i_L(t)$ is the current through the inductor, and $L$ is a constant called the inductance of the inductor. The ideal dynamic elements are, unlike resistors, lossless elements, i.e., they do not dissipate energy but accumulate it. The energy accumulated at a given moment by such an element can be subsequently transferred to the circuit in which the respective element is connected.

A circuit that contains at least one dynamic element is called *a dynamic circuit*. The behavior of dynamic circuits, consisting of independent sources, inductors, capacitors, and resistors, is described by a system of differential equations.

### 2.1 First-order dynamic circuits

A first-order linear circuit contains only one dynamic element (an inductor or a capacitor), other linear circuit elements (resistors, linear controlled sources), and independent sources. The resistive two poles have an equivalent voltage generator (Thevenin) in **Figure 1** [4] and/or an equivalent current generator (Norton) in

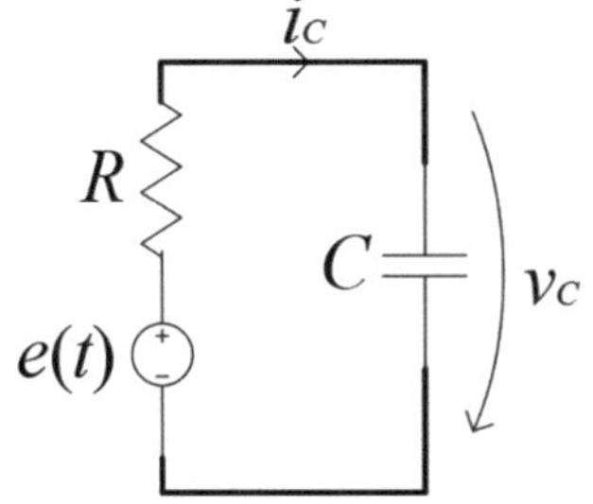


**Figure 1.**
*A Thevenin first-order circuit.*

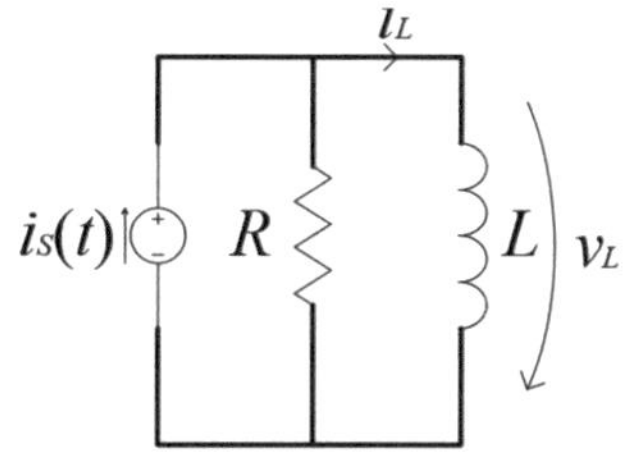


**Figure 2.**
*A Norton first-order circuit.*

**Figure 2** [5] at the input of the dynamic element. With this consideration in mind, it is sufficient to consider only the next two first-order linear circuits.

| Equations of Figure 1's circuits | Equations of Figure 2's circuits |
|---|---|
| $R{\cdot}i_c + v_c = e(t)$ | $\frac{v_L}{R} + i_L = i_s(t)$ |
| $i_c = C{\cdot}\frac{dv_c}{dt}$ | $v_L = L{\cdot}\frac{di_L}{dt}$ |
| $R{\cdot}C{\cdot}\frac{dv_c}{dt} + v_c = e(t)$ | $\frac{L}{R}\cdot\frac{di_L}{dt} + i_L = i_s(t)$ |
| $\frac{du_c}{dt} + \frac{v_c}{R{\cdot}C} = \frac{e(t)}{R{\cdot}C}$ | $\frac{di_L}{dt} + \frac{R}{L}{\cdot}i_L = \frac{R}{L}{\cdot}i_s(t)$ |
| If we note: | |
| $v_c = x$ | $i_L = x$ |
| $\frac{dv_c}{dt} = \dot{x}$ | $\frac{di_L}{dt} = \dot{x}$ |
| $R{\cdot}C = \tau$ | $\frac{L}{R} = \tau$ |
| $e(t) = s(t)$ | $i_s(t) = s(t)$ |
| then | |
| $\dot{x} + \frac{x}{\tau} = \frac{s(t)}{\tau}$ | $\dot{x} + \frac{x}{\tau} = \frac{s(t)}{\tau}$ |

So, a first-order linear circuit satisfies the equation:

$$\dot{x} + \frac{x}{\tau} = \frac{s(t)}{\tau} \tag{1}$$

where $x$ is the state variable of the circuit, $\tau$ is the time constant of the circuit, and $s(t)$ is the parameter of the equivalent independent source.

## 2.2 Dynamic circuits of second order or greater

The aim is to write the state equations of state in normal form $\dot{x} = f(x,t)$. Bringing the equations to this form has two advantages, the qualitative properties of the circuit can be studied more easily, and certain numerical methods for circuit analysis can be applied, formulated to solve a system of differential equations written in this form.

There are two cases. In the first case, the circuit does not contain any loop consisting only of independent or controlled voltage sources and/or capacitors and does not contain any cut-set consisting only of independent or controlled current sources and/or inductors. In this case, the state variables are independent of each other, and we say that *the circuit has no excess state quantities*. In the second case, the circuit does not satisfy the above restrictions, the state variables are not independent of each other, and we say that *the circuit has excess state quantities*.

### *2.2.1 Circuits without excess state quantities*

Any linear dynamic circuit without excess state quantities can be considered as a linear resistive multiport (containing linear resistors and independent sources) with dynamic elements connected at the ports. The dynamic circuit of order $n > 2$ that contains $p$ capacitors and $n - p$ inductors can be represented as follows, in **Figure 3**.

Capacitors are replaced with independent voltage sources and inductors with independent current sources, as in **Figure 4**. If the following notations are made:

$x = [v_1, \ \cdots, \ v_p, \ i_{p+1}, \ \cdots, \ i_n]^t$—vector of state variables,

$y = [i_1, \ \cdots, \ i_p, \ v_{p+1}, \ \cdots, \ v_n]^t$—the vector of the output quantities and

$u = [e_1, \ \cdots, \ e_\alpha, \ i_{\alpha+1}, \ \cdots, \ i_\mu]^t$—the vector of the input quantities (independent sources), according to the superposition theorem [6], the circuit equations become $y = k_0 \cdot x + k_1 \cdot u$, where $k_0$ and $k_1$ are matrices with constant elements.

Using the notation: $\Delta = \text{diag}(C_1, \ \cdots, \ C_p, \ L_{p+1}, \ \cdots \ L_n)$ we obtain $\dot{x} = \Delta^{-1} \cdot y$ and $y = \Delta \cdot \dot{x}$ and the state equations are $\dot{x} = \Delta^{-1} \cdot (k_0 \cdot x + k_1 \cdot u)$ or:

$$\dot{x} = A \cdot x + B \cdot u \qquad (2)$$

where $A$ is called the circuit state matrix.

As an example, for the circuit without excess state quantities in the **Figure 5** [3, 7], the following state equations are obtained.

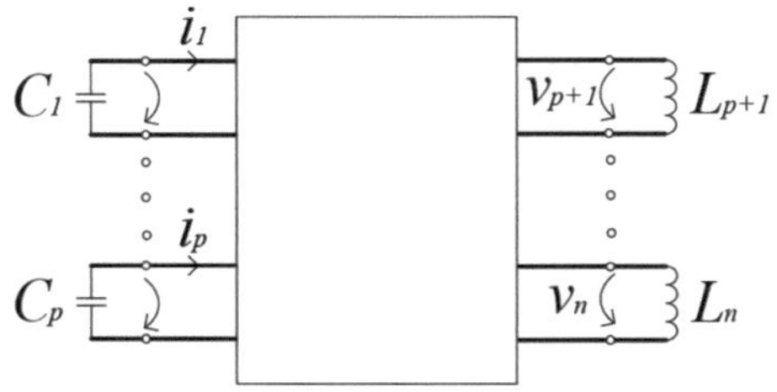


**Figure 3.**
*A p capacitors and n − p inductors dynamic circuit.*

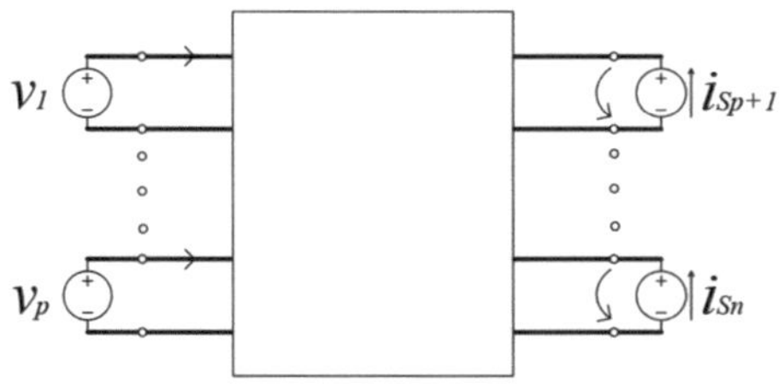


**Figure 4.**
*Capacitors/inductors replaced with independent voltage/current sources.*

*DOI: http://dx.doi.org/10.5772/intechopen.105780*

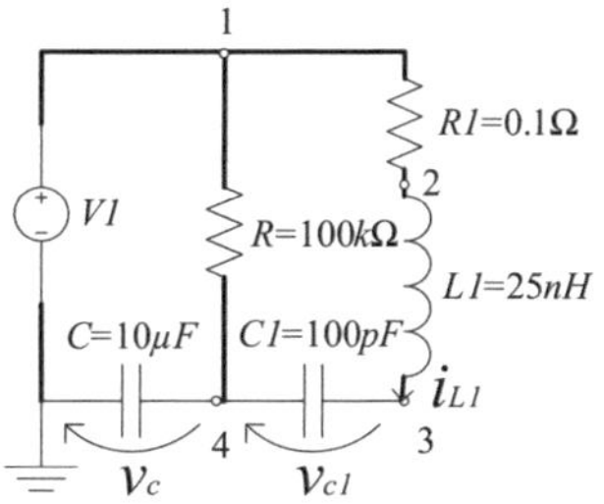


$$\begin{bmatrix} \dot{v}_C \\ \dot{v}_{C_1} \\ \dot{i}_{L_1} \end{bmatrix} = \begin{bmatrix} -\frac{1}{R\cdot C} & 0 & \frac{1}{C} \\ 0 & 0 & \frac{1}{C_1} \\ -\frac{1}{L_1} & -\frac{1}{L_1} & -\frac{R_1}{L_1} \end{bmatrix} \cdot \begin{bmatrix} v_C \\ v_{C_1} \\ i_{L_1} \end{bmatrix} + \begin{bmatrix} -\frac{1}{R\cdot C} \\ 0 \\ -\frac{1}{L_1} \end{bmatrix} \cdot e(t)$$

**Figure 5.**
*Circuit without excess state quantities.*

### 2.2.2 Circuits with excess state quantities

In this case, there is at least one loop consisting only of capacitors and voltage sources or a cut-set consisting only of inductors and current sources. This means that there will be at least one capacitor that cannot be placed in the tree or an inductor that cannot be placed in the co-tree. The normal tree therefore contains all voltage sources and as many capacitors as possible whose voltages are independent state quantities. The other capacitors will be contained in the normal co-tree, and their voltages are excess state quantities. The normal co-tree contains all current sources and as many inductors as possible whose currents are independent state quantities. The other inductors are contained in the normal tree and their currents are excess state quantities. We consider the circuit as a resistive multiport with dynamic elements connected to the ports. According to the substitution theorem, the capacitors and inductors in the tree are replaced with voltage sources and the capacitors and inductors in the co-tree with current sources. The result is the circuit in **Figure 6**, in which for simplicity, only one source was represented for each category of element (tree capacitors, tree inductors, co-tree inductors, co-tree capacitors). The second size index represents tree $(t)$ or co-tree $(c)$.

Using the notation : $x = [v_{Ct}, \quad i_{Lc}]^t, x^* = [v_{Cc}, \quad i_{Lt}]^t, y = [i_{Ct}, \quad u_{Lc}]^t, y^* = [i_{Cc}, \quad v_{Lt}]^t$.

The resistive circuit being linear, according to the superposition theorem we obtain:

$$y = k_0 \cdot x + k_1 \cdot \mu_S + k_2 \cdot y^* \tag{3}$$

where $k_0$, $k_1$, and $k_2$ are matrices with constant parameters. The operating equations of the dynamic elements can be written:

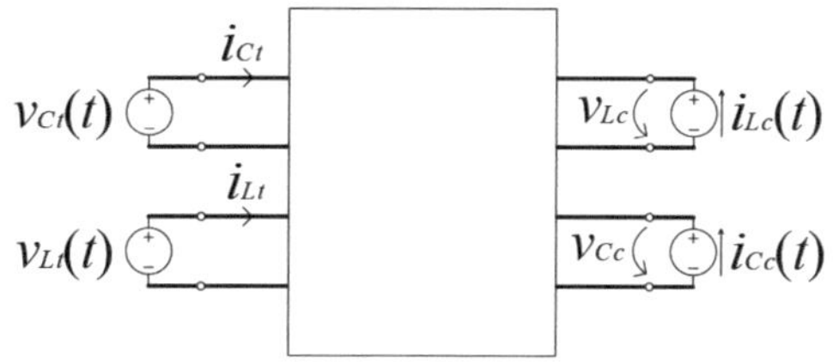


**Figure 6.**
*The capacitors and inductors in the tree replaced with voltage sources, and the capacitors and inductors in the co-tree with current sources.*

$$y = -\Delta \cdot \dot{x} \text{ where } \Delta = \mathrm{diag}(\, C_t, \quad L_c\,),$$
$$y^* = -\Delta^* \cdot \dot{x}^* \text{ where } \Delta^* = \mathrm{diag}(\, C_c, \quad L_t\,).$$

The excess state quantities can be expressed, with the help of the Kirchhoff's laws, depending on the independent state quantities and the parameters of some independent sources: $x^* = k_3 \cdot x + k_4 \cdot \mu_S$ and we obtain:

$$y^* = -\Delta^* \cdot k_3 \cdot \dot{x} - \Delta^* \cdot k_4 \cdot \dot{\mu}_S \tag{4}$$

But:

$$\dot{x} = -\Delta^{-1} \cdot y = -\Delta^{-1} \cdot [k_0 \cdot x + k_1 \cdot \mu_S - k_2 \cdot (\Delta^* \cdot k_3 \cdot \dot{x} + \Delta^* \cdot k_4 \cdot \dot{\mu}_S)] \tag{5}$$

so, the last relationship can be written as:

$$\dot{x} = A \cdot x + B \cdot \mu_S + B^* \cdot \dot{\mu}_S \tag{6}$$

where $A$ is the state matrix of the circuit.

As an example, for the circuit with excess state quantities (capacitor loop and inductor cut-set circuit) and without sources for simplicity, in **Figure 7** [8, 9], the state equations are obtained:

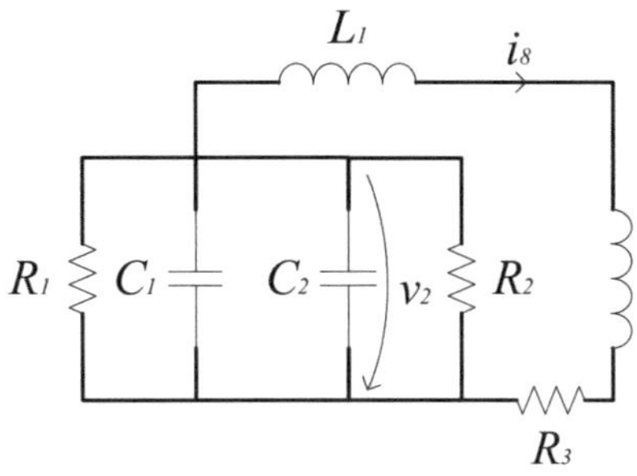


$$\begin{bmatrix} \dot{v}_2 \\ \dot{i}_8 \end{bmatrix} = \begin{bmatrix} -\dfrac{R_1 + R_2}{(C_1 + C_2) \cdot R_1 \cdot R_2} & -\dfrac{1}{C_1 + C_2} \\ \dfrac{1}{L_1 + L_2} & \dfrac{R_3}{L_1 + L_2} \end{bmatrix} \cdot \begin{bmatrix} v_2 \\ i_8 \end{bmatrix}$$

**Figure 7.**
*Circuit with excess state quantities.*

## 2.3 Dynamic circuits with resistive nonlinearities

The methods presented above for writing state equations can also be used and in the case of dynamic circuits with resistive nonlinearities, with only a few changes. A first approach is to approximate the nonlinear characteristic of the resistor with piece-wise linear characteristic. The state equations for the nonlinear circuit can be written separately for each linear portion of the piece-wise linear characteristic. For example, for the rectifier with a diode and the *RC* load in **Figure 8**, we have two states. One in which the diode conducts and is equivalent to a very low value resistor, ideal 0Ω as in **Figure 9** (state 1) and the other in which the diode does not conduct and is equivalent to a very high value resistor, ideal ∞Ω as in **Figure 10** (state 2).

From the equations obtained separately for the two states, the following state equation can be written:

$$\frac{dv_c(t)}{dt} = \left( -\frac{1-D}{R_E \cdot C} - \frac{1}{R \cdot C} \right) \cdot v_c(t) - \frac{1-D}{R_E \cdot C} \cdot e(t) \tag{7}$$

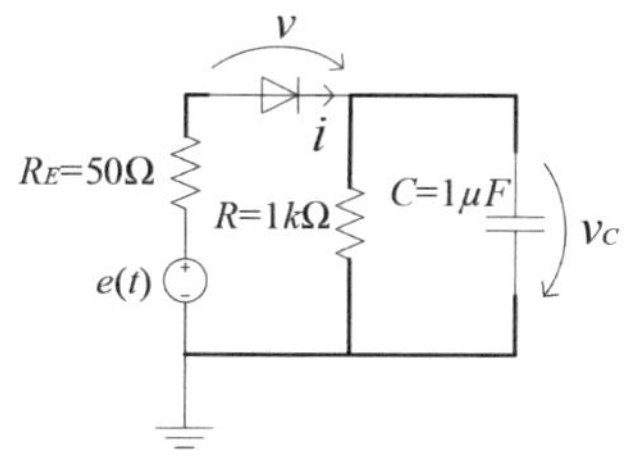


**Figure 8.**
*A diode rectifier with RC load.*

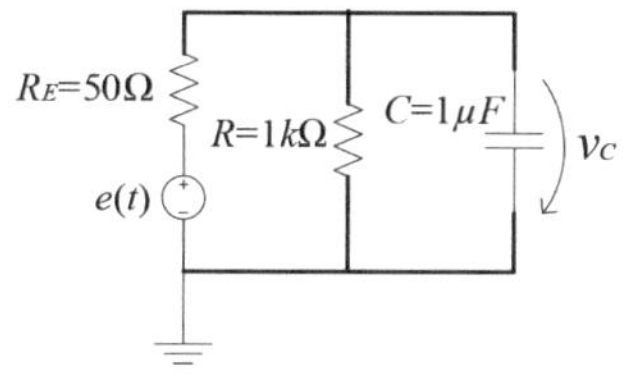


**Figure 9.**
*Diode conducts (state 1).*

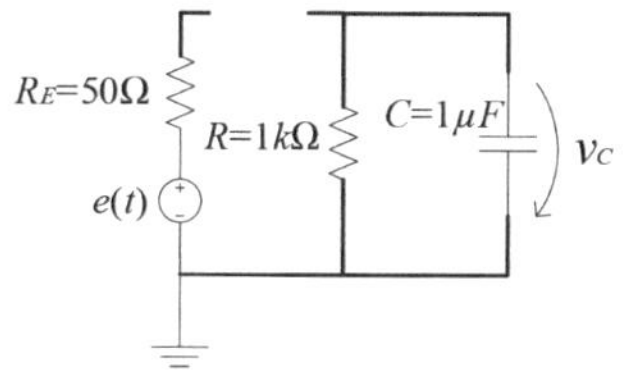


**Figure 10.**
*Diode does not conduct (state 2).*

Where $e(t) = 5 \cdot sin\ (2\pi f \cdot (t + t_0))\ [V]$ and $f = 10kHz$.

When $D = 0$ the equation for state 1 is obtained, and when $D = 1$ the equation for state 2 is obtained. $D$ acts as a closed-open switch over a well-defined time interval depending on the state of the diode. If the rectifier diode conducts, $D$ acts as an open switch, and if the diode does not conduct, $D$ acts as a closed switch [10].

Another approach to writing the state equations for dynamic circuits with resistive nonlinearities is to replace in the state equations the nonlinear resistance with the function that describes the nonlinearity (**Figure 11**). In the example of a rectifier with a diode and $RC$ load, the diode can be modeled as a nonlinear piece-wise linear function: $R_D = 0.1\Omega$ if the voltage at its terminals is positive $v > 0$, and $R_D = 10M\Omega$ if the voltage at its terminals is negative $v < 0$, **Figure 12**.

The following state equation is obtained:

$$\frac{dv_c(t)}{dt} = -\frac{R_D + R_E + R}{(R_D + R_E) \cdot R \cdot C} \cdot v_c(t) + \frac{1}{(R_D + R_E) \cdot C} \cdot e(t) \tag{8}$$

As the handling of functions is more difficult than that of symbols, the second approach is suitable for small circuits and the first for larger circuits.

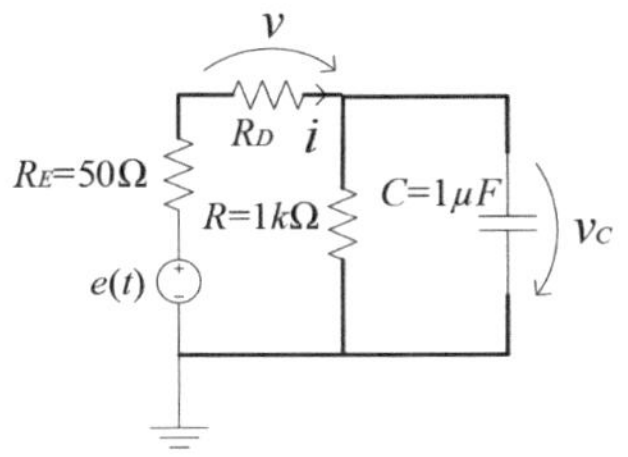


**Figure 11.**
*A rectifier with RC load; diode replaced with a nonlinear resistance.*

$$R_D = \begin{cases} 0.1\Omega \text{ for } v > 0 \\ 10M\Omega \text{ for } v < 0 \end{cases}$$

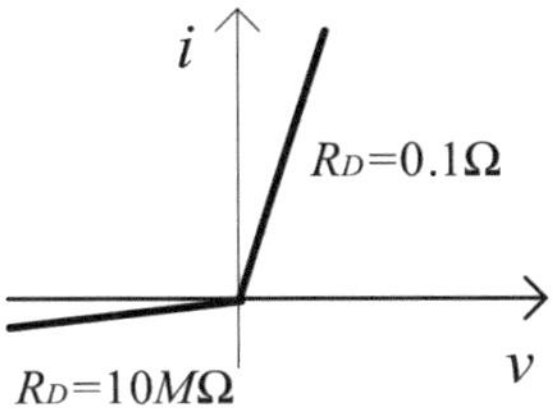


**Figure 12.**
*The nonlinear resistance function.*

## 3. Qualitative behavior of dynamic circuits

### 3.1 First-order dynamic circuits

In the previous chapter, it has been shown that a linear circuit of the first order satisfies the equation:

$$\dot{x} + \frac{x}{\tau} = \frac{s(t)}{\tau} \tag{9}$$

where $x$ is the state variable of the circuit, $\tau$ is the time constant of the circuit, and $s(t)$ is the parameter of the equivalent independent source.

The solution of a first-order linear circuit with independent direct current sources consists of two terms: the solution of the homogeneous equation $x_v$ and a particular solution $x_p$:

$$x = x_v + x_p \tag{10}$$

The homogeneous equation is $\dot{x}_v + \frac{x_v}{\tau} = 0$. This can be solved using the separation of variables method:

$$\dot{x}_v = -\frac{x_v}{\tau}$$

$$\frac{dx_v}{dt} = -\frac{x_v}{\tau}$$

$$\frac{dx_v}{x_v} = -\frac{dt}{\tau}$$

$$\int_{x_0}^{x_v} \frac{dx_v}{x_v} = -\int_{t_0}^{t} \frac{dt}{\tau}$$

$$ln\,(x_v) - ln\,(x_0) = -\frac{(t - t_0)}{\tau}$$

$$ln\,(x_v) = C_1 - \frac{(t - t_0)}{\tau}$$

$$x_v = e^{C_1} \cdot e^{-\frac{(t-t_0)}{\tau}} = C_2 \cdot e^{-\frac{(t-t_0)}{\tau}}$$

The particular solution is a constant, in the form of free term $x_p = C_3$, so:

$$x(t) = C_2 \cdot e^{-\frac{(t-t_0)}{\tau}} + C_3$$

The constant $C_3$ is calculated by replacing $t = t_\infty$:

$$x(t_\infty) = C_2 \cdot e^{-\infty} + C_3 = 0 + C_3 \text{ so } C_3 = x(t_\infty)$$

The solution is determined only if the initial condition is known $x(t_0)$. Replacing $t = t_0$ we obtain:

$$x(t_0) = C_2 \cdot e^0 + x(t_\infty) = C_2 + x(t_\infty) \text{ so } C_2 = x(t_0) - x(t_\infty) \text{ and results :}$$

$$x(t) = [x(t_0) - x(t_\infty)] \cdot e^{-\frac{(t-t_0)}{\tau}} + x(t_\infty) \tag{11}$$

We distinguish two cases in which the behavior of the solution is different: $\tau > 0$ and $\tau < 0$. If $\tau < 0$, when $t \to \infty$, $x(t)$ decreases exponentially over time and the solution *tends to equilibrium* (**Figure 13**). If $\tau < 0$, when $t \to \infty$, $x(t)$ increases exponentially over time and the solution tends to an *infinite value* (**Figure 14**).

### 3.2 Dynamic circuits of second order or greater

The qualitative behavior of the circuit is determined by the eigenvalues of the state matrix $A$. These can be calculated as roots of the equation:

$$det(A - \lambda \cdot I) = 0 \tag{12}$$

where $I$ is the unit matrix of the same order as the state matrix $A$.

First case: The state matrix $A$ has real and distinct eigenvalues. There are three possibilities:

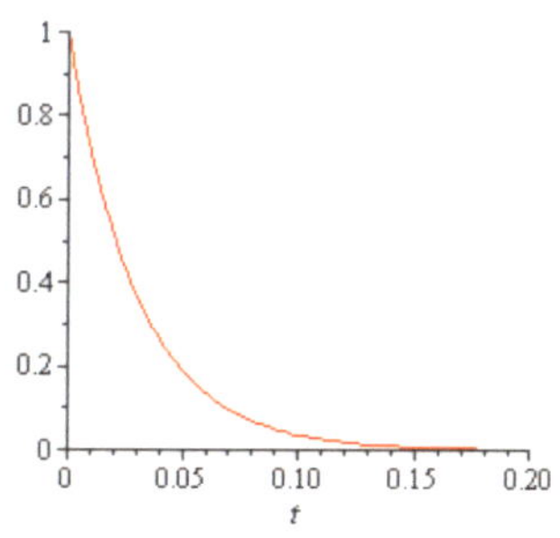


**Figure 13.**
*The solution decreases exponentially.*

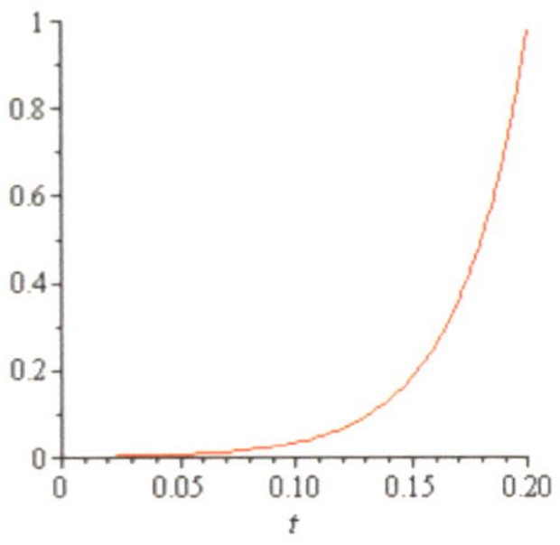


**Figure 14.**
*The solution increases exponentially.*

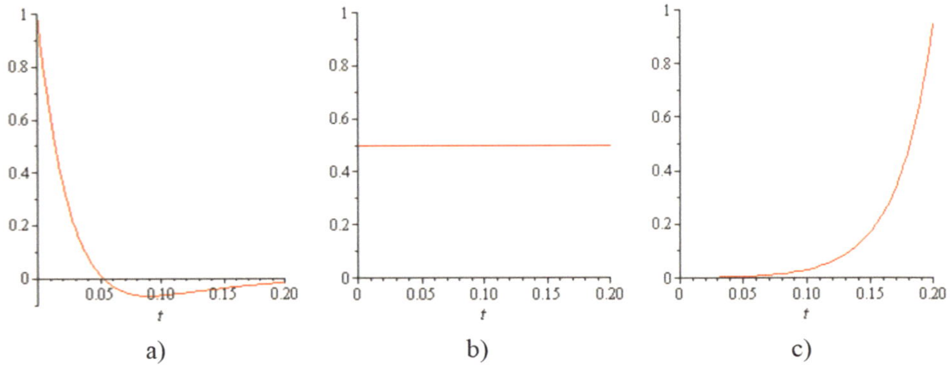


**Figure 15.**
*Circuit response for real and distinct eigenvalues (a) stable response; (b) constant response; and (c) unstable response.*

a. the state matrix $A$ has negative eigenvalues. In this case the solution moves towards the *steady state* when $t \to \infty$ (**Figure 15a**).

b. the state matrix $A$ has null eigenvalues. In this case the solution is a constant, the *steady state*, when $t \to \infty$ as well as when $t \to -\infty$ (**Figure 15b**).

c. the state matrix $A$ has positive eigenvalues. In this case the solution moves towards the *steady state* when $t \to -\infty$ (**Figure 15c**).

Second case: The state matrix $A$ has complex conjugated eigenvalues. In this case, there are also three possibilities:

a. state matrix $A$ has eigenvalues with a negative real part. In this case, the solution contains damped harmonic components that move toward the *steady state* when $t \to \infty$ (**Figure 16a**).

b. The state matrix $A$ has eigenvalues with null real part. In this case, the solution contains undamped harmonic components when $t \to \infty$ as well as when $t \to -\infty$ (**Figure 16b**).

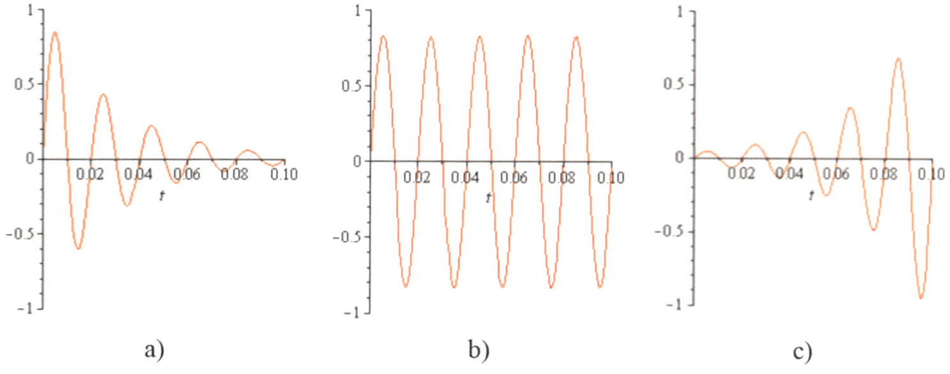


**Figure 16.**
*Circuit response for complex conjugated eigenvalues (a) stable response; (b) constant response; and (c) unstable response.*

c. The state matrix $A$ has eigenvalues with a positive real. In this case, the solution contains damped harmonic components that go toward the *steady state* when $t \to -\infty$ (**Figure 16c**).

It can be observed that circuits that have the state matrix $A$ with negative real or complex conjugated eigenvalues with the negative real part are stable. If at least one real eigenvalue is positive or a pair of complex conjugated eigenvalues has a positive real part, they are unstable.

For example, for the circuit in paragraph 2.2.1 (**Figure 5**), by replacing the numerical values for the circuit elements in the state matrix, the following eigenvalues are obtained:

$$det(A - \lambda \cdot I) = det \begin{pmatrix} -1-\lambda & 0 & 10^5 \\ 0 & -\lambda & 10^{10} \\ -4 \cdot 10^7 & -4 \cdot 10^7 & -4 \cdot 10^6 - \lambda \end{pmatrix} = 0$$

So:

$$-\lambda^3 - 4 \cdot 10^6 \cdot \lambda^2 - 4 \cdot 10^{17} \cdot \lambda - 4 \cdot 10^{17} = 0 \to \begin{cases} \lambda_1 = -1 \\ \lambda_2 = -2 \cdot 10^6 + 6.3246 \cdot 10^8 \cdot i \\ \lambda_3 = -2 \cdot 10^6 - 6.3246 \cdot 10^8 \cdot i \end{cases}$$

It is observed that all eigenvalues have a negative real part, so the circuit is stable.

## 4. Analytical solving of dynamic circuits equations

With the help of the Laplace transform, it is possible to construct a system of algebraic equations in the complex variable $s$ domain, which corresponds to a system of linear differential equations in the time domain. The use of algebraic equations instead of differential equations shows evident benefits in the study of electrical circuits.

Although section 2 shows how to write state equations in normal form $\dot{x} = f(x, t)$, this method is cumbersome for a human operator even for circuits with a small number of dynamic elements. For this reason, in the analysis of electrical circuits, it is preferred to apply the Laplace transform first to the equations that describe the operation of the circuit elements and then to write the circuit equations.

## 4.1 Circuit analysis with Laplace transform

The study of the time-varying regime can be performed if the initial conditions ($i_k(0_-)$ inductor currents and $v_k(0_-)$ capacitor voltages) at $t = 0_-$ are known. It is considered that the independent sources are connected in the circuit at $t = 0_-$. In this case, all voltages and all currents are original functions. For example, a direct current source (voltage or current) having $E = ct$ or $I_S = ct$, being connected at $t = 0_-$ has $e(t) = E{\cdot}u(t)$ or $i_S(t) = I_S{\cdot}u(t)$. Thus, the quantities $e(t)$ and $i_S(t)$ become original functions due to the presence of the factor $u(t)$. For such a circuit there are Laplace images of voltages and currents: $I_k(s) = L\{i_k(t)\}$ and $V_k(s) = L\{v_k(t)\}$. Laplace image computation is called operational computation. Applying the linearity property of the Laplace transform, Kirchhoff's Laws $\sum_N i_k(t) = 0$ and $\sum_B v_k(t) = 0$ can be written in the complex variable $s$ domain: $\sum_N I_k(s) = 0$ and $\sum_B V_k(s) = 0$.

According to the linearity and derivation properties of the original function of the Laplace transform, the differential equations that express the connections between $v_k(t)$ and $i_k(t)$ for the circuit elements, correspond to algebraic equations that express the connections between $V_k(s) = L\{v_k(t)\}$ and $I_k(s) = L\{i_k(t)\}$ [11].

## 4.2 Equivalent operational circuits

A circuit in which currents and voltages are Laplace images is called the equivalent operational circuit. Below are the most common circuit elements and their Laplace images [11].

a. The ideal direct current voltage (or current) source

| | | |
|---|---|---|
| 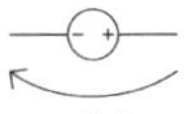 $e(t)$ | $\rightarrow$ |  $E(s)$ |
| $e(t) = E$ | | $E(s) = \frac{E}{s}$ |

b. The ideal alternating current voltage (or current) source

| | | |
|---|---|---|
| 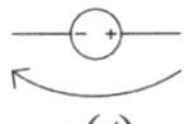 $e(t)$ | $\rightarrow$ |  $E(s)$ |
| $e(t) = E \cdot sin\,(\omega \cdot t + \phi)$ | | $E(s) = E \cdot \frac{\omega \cdot cos\,(\phi) + s \cdot sin\,(\phi)}{\omega^2 + s^2}$ |

c. The resistor

The ideal resistor has the operation equation $v(t) = R \cdot i(t)$ and if $V(s) = L\{\,v(t)\}$, $I(s) = L\{i(t)\}$ then $V(s) = R \cdot I(s)$. The factor multiplied with $I(s)$ to get $V(s)$ is called the operational impedance $\left(Z(s) = \frac{V(s)}{I(s)}\right)$. For the ideal resistor $Z_R(s) = R$. The circuit corresponding to this relationship is shown below:

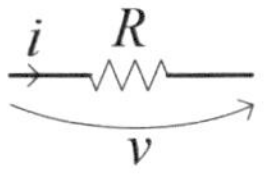


→

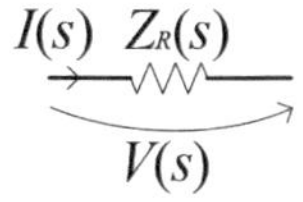


| | |
|---|---|
| $v(t) = R \cdot i(t)$ | $V(s) = R \cdot I(s)$ |

d. The inductor

For the ideal inductor, according to the derivation property of the original function, the equation $v(t) = L \cdot \frac{di(t)}{dt}$ transforms to:

$$V(s) = L \cdot [s \cdot I(s) - i(0_-)] = s \cdot L \cdot I(s) - L \cdot i(0_-) \tag{13}$$

where $i(0_-)$ is the initial condition at the time $t = 0_-$ for the current through the inductor. The equivalent operational circuit is:

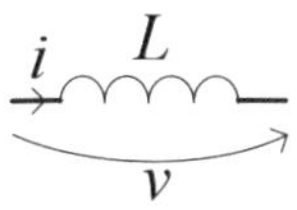


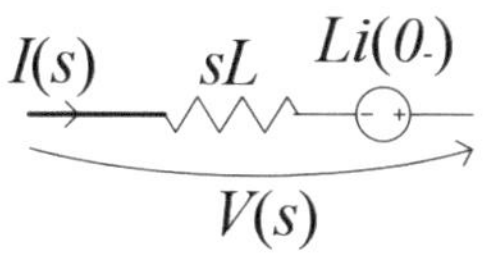


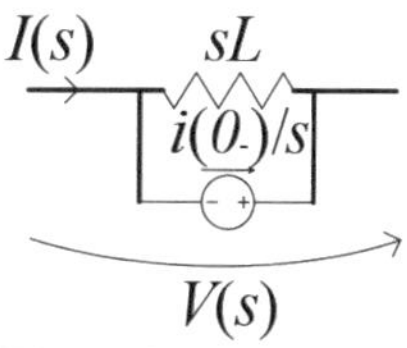


| | | |
|---|---|---|
| $v(t) = L \cdot \frac{di(t)}{dt}$ | $V(s) = s \cdot L \cdot I(s) - L \cdot i(0_-)$ | $I(s) = \frac{V(s)}{s \cdot L} + \frac{i(0_-)}{s}$ |

e. The capacitor

Similar to the inductor, according to the derivation property of the original function, for the ideal capacitor, the relationship $i(t) = C \cdot \frac{dv(t)}{dt}$ transforms to:

$$I(s) = C \cdot [s \cdot V(s) - v(0_-)] = \frac{V(s)}{\frac{1}{s \cdot C}} - C \cdot v(0_-) \tag{14}$$

where $v(0_-)$ is the initial condition at the $t = 0_-$ for the voltage drop on the capacitor. The equivalent operational circuit is:

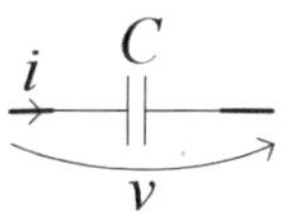


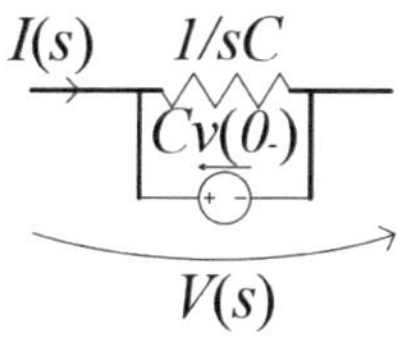


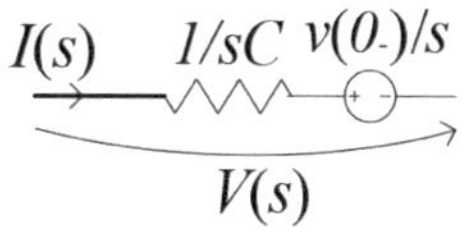


| | | |
|---|---|---|
| $i(t) = C \cdot \frac{dv(t)}{dt}$ | $I(s) = \frac{V(s)}{\frac{1}{s \cdot C}} - C \cdot v(0_-)$ | $V(s) = \frac{I(s)}{s \cdot C} + \frac{v(0_-)}{s}$ |

### 4.3 Response computation of linear dynamic circuits

All theorems and analysis methods valid for direct or alternating current circuits are valid for the equivalent operational circuits: equivalent generator theorems, superposition theorem, the two-port representation theorems, Kirchhoff's laws analysis, nodal analysis, mesh analysis, etc.

The Laplace transform analysis algorithm of a linear circuit in time-varying regime consists of:

1. determine the initial conditions for the inductors and capacitors in the circuit, $i_L(0_-)$ and $v_C(0_-)$;

2. build the circuit with operational sources and operational impedances using the operational equivalent circuits;

3. write the equations and solve for the unknowns $V_K(s)$ and $I_K(s)$;

4. determine the analytical solution, the original functions $v_k(t)$ and $i_k(t)$ as the inverse Laplace transform (using Heaviside's theorems).

If the circuit has more than four dynamic elements, determining the circuit response using analytical methods requires substantial effort for a human operator. In this case, a software product such as Mathematica [12] or Maple [13] capable of performing symbolic computations can be used.

To illustrate the above algorithm, consider the following example, the circuit in **Figure 17** [14]. The current through the inductor $i_L(t)$ and the voltage at the terminals of the capacitor $v_C(t)$ in the transient mode that occurs after the switch $K$ *is closed* at the time $t = 0$ in the circuit in the figure below are required. We know $R = 1\Omega$, $C = 1/3F$, $L = 3/2H$, $E = 6V$, $i_L(0_-) = 3A$ and $v_C(0_-) = 3V$.

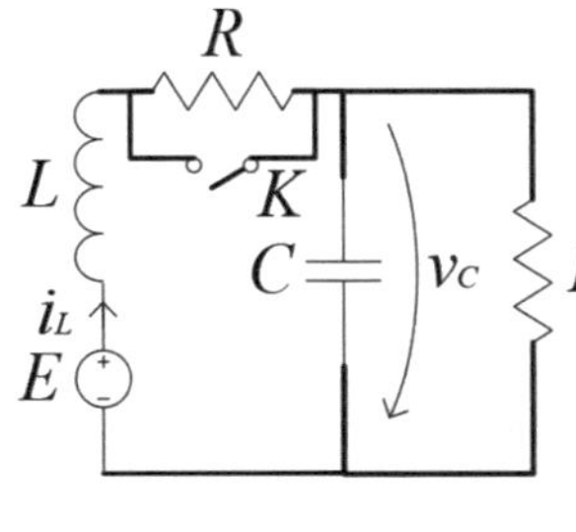


**Figure 17.**
*Dynamic linear circuit in transient mode.*

Using the equivalent operational circuits (**Figure 18**), we obtain:

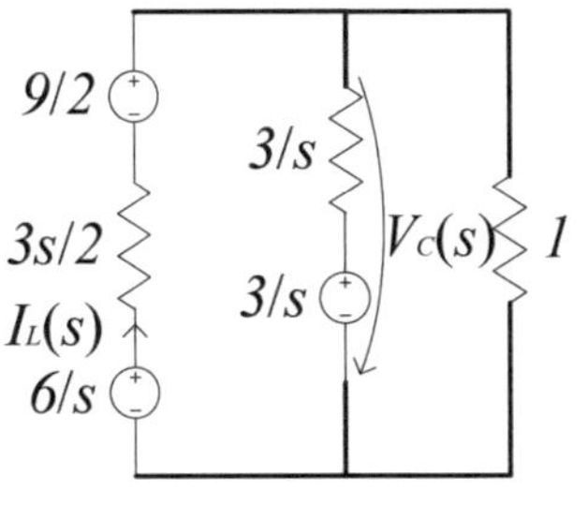


**Figure 18.**
*The equivalent operational circuit.*

Writing and solving the circuit equations results in the inductor current:

$$I_L(s) = \frac{3 \cdot s^2 + 11 \cdot s + 12}{s \cdot (s+1) \cdot (s+2)}$$

and capacitor voltage:

$$V_C(s) = \frac{3 \cdot s^2 + 9 \cdot s + 18}{s \cdot (s+1) \cdot (s+2)}$$

To make it easier to calculate the transient solution, we breakdown the expression into simple fractions:

$$I_L(s) = \frac{6}{s} + \frac{-4}{s+1} + \frac{1}{s+2} \text{ and } V_C(s) = \frac{6}{s} + \frac{-6}{s+1} + \frac{3}{s+2}$$

So:

$$i_L(t) = L^{-1}\{I_L(s)\} = 6 - 4 \cdot e^{-t} + e^{-2 \cdot t} \ [A]$$

and

$$v_C(t) = L^{-1}\{V_C(s)\} = 6 - 6 \cdot e^{-t} + 3 \cdot e^{-2 \cdot t} \ [V]$$

### 4.4 Response computation of the dynamic circuits with resistive nonlinearities

Solving systems of differential equations by a human operator is difficult even in the case of linear circuits and even more so in the case of nonlinear ones. To solve them, it is preferred to use programs capable of performing symbolic computations, such as Mathematica, Maple, etc. To solve the circuit state equation obtained by the second approach in paragraph 2.2:

$$\frac{dv_c(t)}{dt} = -\frac{R_D + R_E + R}{(R_D + R_E) \cdot R \cdot C} \cdot v_c(t) + \frac{1}{(R_D + R_E) \cdot C} \cdot e(t) \qquad (15)$$

where: $R_D = \begin{cases} 0.1\Omega \text{ for } v > 0 \\ 10M\Omega \text{ for } v < 0 \end{cases}$, $e(t) = 5 \cdot sin\,(2\pi f \cdot (t + t_0))$ $[V]$ and $f = 10kHz$, it is possible to proceed in this way. The above equation is solved separately for each linear region of the nonlinear characteristic for $R_D$. When $R_D$ value changes, the initial condition $v_c(t_0) = v_{c0}$ is considered as the value of the voltage $v_c(t)$ obtained at the end of the previous interval. These calculations were made in the Maple program for two periods of the source $e(t)$, with the following solution obtained by concatenating the results (**Figure 19**):

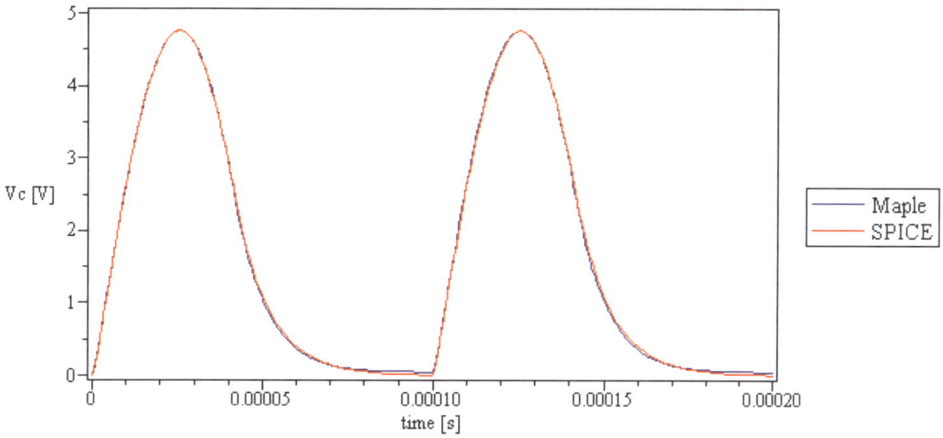


**Figure 19.**
*The analytical solution (Maple) and the numerical solution (SPICE).*

As a verification of this method, we can compare the results with those obtained by using the SPICE program [15], in the same time frame. It is observed that the analytical solution obtained with Maple is almost identical to numerical solution obtained with SPICE.

## 4.5 Transfer functions

Consider a circuit with constant parameters and a unique solution. We assume that we have only one independent source and initial conditions equal to zero. The transfer function from gate $i$ to gate $j$ is defined as the ratio between the output quantity on side $j$ and the input quantity of side $i$ (**Figure 20**). The output quantity can be a voltage or a current, and the input quantity can be the electromotive voltage of an independent voltage source or the current of an independent current source.

Generally, a circuit function also called a transfer function is:

$$H(s) = \frac{P(s)}{Q(s)} \tag{16}$$

The roots of $Q(s)$ are called *the poles of* $H(s)$, and the roots of $P(s)$ are called the *zeros of* $H(s)$. The dynamic behavior of the network depends upon the location of the poles and zeros on the network function curve [16]. In general, one can graphically deduce the magnitude and phase curve of any network function from the location of its poles and zeros. The poles of $H(s)$ are the circuit natural frequencies. Not all natural frequencies are the poles of any function of that circuit because certain common expressions that appear in both $Q(s)$ and $P(s)$ can disappear by simplification.

The poles of the transfer function determine the stability of the circuit, similarly to the eigenvalues of the state matrix presented in paragraph 3. If all the poles have a negative real part, the circuit is stable. If at least one pole has a positive real part, the circuit is unstable.

HSPICE uses the Muller method to calculate the roots of polynomials $P(s)$ and $Q(s)$. This method approximates the polynomial with a quadratic equation that fits through

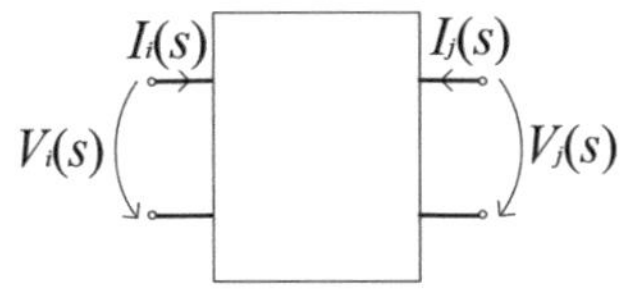


**Figure 20.**
*Circuit with one input (i) and one output (j).*

| Poles | Real | Imaginary |
|---|---|---|
| 1 | $-9.9999 \cdot 10^{-1}$ | 0 |
| 2 | $-2 \cdot 10^{6}$ | $+6.3246 \cdot 10^{8}$ |
| 3 | $-2 \cdot 10^{6}$ | $-6.3246 \cdot 10^{8}$ |

**Table 1.**
*The poles obtained with HSPICE for the circuit in* ***Figure 5****.*

three points in the vicinity of a root. Successive iterations toward a particular root are obtained by finding the nearest root of a quadratic equation whose curve passes through the last three points [16]. Pole/zero analysis results are based on the circuit's DC operating point, so the operating point solution must be accurate [16].

For the circuit in paragraph 2.2.1, **Figure 5**, the poles in **Table 1** are obtained using the HSPICE program.

It is observed that they are identical with the eigenvalues of the state matrix, calculated in paragraph 3.2.

## 5. Numerical approach of solving dynamic circuits equations

We aim to solve a system of state equations written in normal form $\dot{x} = f(x,t)$. Even for linear circuits whose equations have an analytical solution, the use of numerical methods is preferred because even for a second-order circuit, the analytical calculations are quite complicated to be performed efficiently by a human operator.

Automatic solving of analytical solutions requires considerable computational effort, as computers are designed to operate with numbers rather than symbols. For this purpose, specialized software such as Mathematica or Maple that performs analytical calculations can be used.

### 5.1 Numerical integration methods used in circuit simulation

Any numerical method starts from the initial condition $x(t_0)$ and determines successively:

$$x(t_0 + h), x(t_0 + 2h), \ldots, \tag{17}$$

where $h$ is the time step.

The simplest numerical integration methods used in circuit simulation programs are [17]:

#### *5.1.1 The forward Euler method (FE)*

Expanding $x(t_{k+1})$ in Taylor series in the vicinity of the point $t_k$ we obtain:

$$x(t_{k+1}) = x(t_k) + \left.\frac{dx}{dt}\right|_{t_k} \cdot \frac{(t_{k+1} - t_k)}{1!} + \left.\frac{d^2x}{dt^2}\right|_{t_k} \cdot \frac{(t_{k+1} - t_k)^2}{2!} + \ldots \tag{18}$$

Using the notation $x(t_k) = x_k$, neglecting the higher-order terms, and taking into account the fact that $\frac{dx}{dt} = \dot{x} = f(x,t)$, we obtain:

$$x_{k+1} = x_k + h \cdot \dot{x}_k \text{ or } x_{k+1} = x_k + h \cdot f(x_k) \tag{19}$$

where $h = t_{k+1} - t_k$ (time step).

The graphical interpretation of this method of numerical integration is presented below in **Figure 21**.

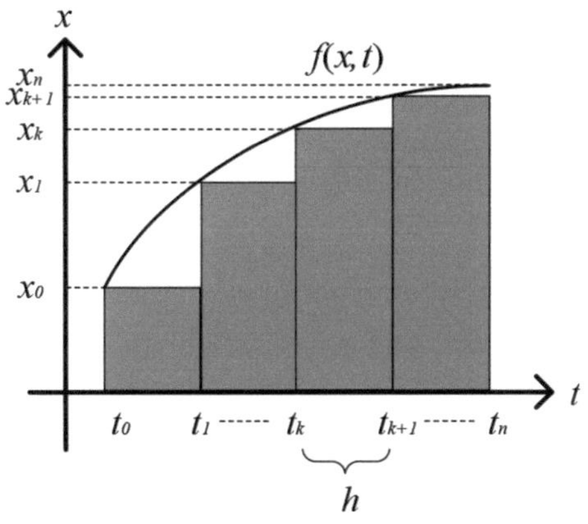


**Figure 21.**
*The forward Euler method (FE) graphical interpretation.*

### *5.1.2 The backward Euler method (BE)*

Similarly, expanding $x(t_k)$ in Taylor series in the vicinity of the point $t_{k+1}$ results:

$$x(t_k) = x(t_{k+1}) + \left.\frac{dx}{dt}\right|_{t_{k+1}} \cdot \frac{(t_k - t_{k+1})}{1!} + \left.\frac{d^2x}{dt^2}\right|_{t_{k+1}} \cdot \frac{(t_k - t_{k+1})^2}{2!} + \ldots \tag{20}$$

With the above notations we obtain:

$$x_{k+1} = x_k + h \cdot \dot{x}_{k+1} \text{ or } x_{k+1} = x_k + h \cdot f(x_{k+1}) \tag{21}$$

In **Figure 22** is presented the graphical interpretation of this numerical integration method.

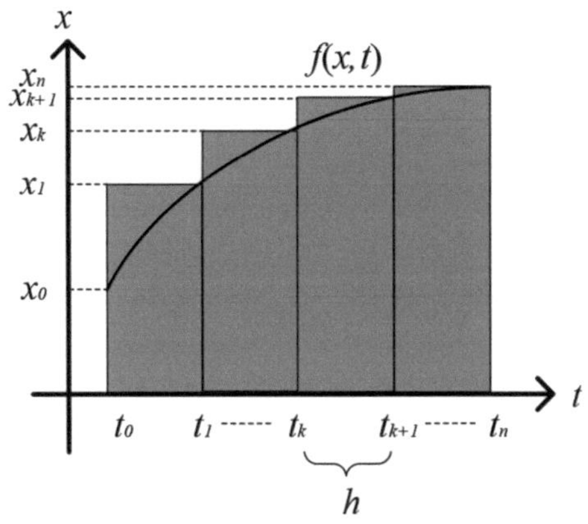


**Figure 22.**
*The backward Euler method (BE) graphical interpretation.*

### *5.1.3 The trapezoidal rule (TR)*

From the graphic interpretation of the two numerical integration methods presented above, it is observed that for the same function, a method approximates the

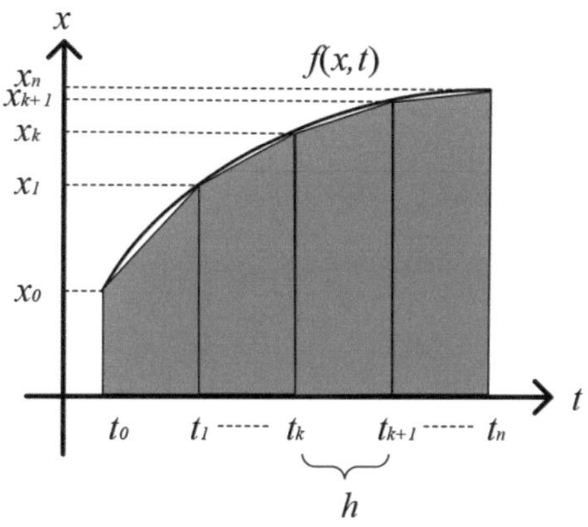


**Figure 23.**
*The trapezoidal rule (TR) graphical interpretation.*

integral with a value smaller and the other with a bigger value. This drawback can be overcome by using the trapezoidal rule.

Adding and dividing by two the two formulas obtained for $x_{k+1}$, the one obtained with the forward Euler method and the one obtained with the backward Euler method, the formula for the trapezoidal rule is obtained:

$$\left.\begin{array}{l} x_{k+1} = x_k + h \cdot \dot{x}_k \\ x_{k+1} = x_k + h \cdot \dot{x}_{k+1} \end{array}\right\} \Rightarrow 2 \cdot x_{k+1} = 2 \cdot x_k + h \cdot (\dot{x}_k + \dot{x}_{k+1})$$

or

$$x_{k+1} = x_k + \frac{h}{2} \cdot (\dot{x}_k + \dot{x}_{k+1}) \tag{22}$$

The graphical interpretation of the trapezoidal rule is presented below, in **Figure 23**.

It is easy to see that the forward Euler method, in which $x(t_{k+1})$ is determined from $x(t_k)$, is an explicit method, while the backward Euler method and the trapezoidal rule are implicit methods.

The solutions obtained by numerical integration being approximate, the errors introduced at each step are calculated (*local error* and *global error* of the method). Local error at the time $t = t_{n+1}$ is $\varepsilon_n = x_{exact}(t_{n+1}) - x_{approx}(t_{n+1})$ where $x_{exact}(t_{n+1})$ was calculated starting from the $x(t_n)$ approximate calculation. The total error at $t = t_{n+1}$ is $\varepsilon_n = x_{exact}(t_{n+1}) - x_{approx}(t_{n+1})$, where $x_{exact}(t_{n+1})$ was calculated from the initial $x(0)$.

The numerical integration method for which the total error decreases as time passes is a *stable method*. A method that does not have this property is numerically *unstable*, even if the local error is small and decreases over time.

In a stable method, the size of the time step is limited only by the imposed local error, which depends on the studied problem. Working with very low values of $h$ leads to a large number of time steps required to determine the solution that take up an unjustifiably high computation time.

### *5.1.4 The gear methods (G)*

For circuits with eigenvalues that differ by a few orders of magnitude ("stiff") the numerical methods presented above do not give correct results. In this case, special gear methods are used. The relationships that define the second to sixth-order gear methods are:

Second-order gear: $x_{k+1} = \frac{3}{4}x_k - \frac{1}{3}x_{k-1} + h\frac{2}{3} \cdot \dot{x}_{k+1}$

Third-order gear: $x_{k+1} = \frac{18}{11}x_k - \frac{9}{11}x_{k-1} + \frac{2}{11} \cdot x_{k-2} + h\frac{6}{11} \cdot \dot{x}_{k+1}$

Fourth-order gear 4: $x_{k+1} = \frac{48}{25}x_k - \frac{36}{25}x_{k-1} + \frac{16}{25} \cdot x_{k-2} - \frac{3}{25} \cdot x_{k-3} + h\frac{12}{25} \cdot \dot{x}_{k+1}$

Fifth-order gear 5: $x_{k+1} = \frac{300}{137}x_k - \frac{300}{137}x_{k-1} + \frac{200}{137} \cdot x_{k-2} - \frac{75}{137} \cdot x_{k-3} + \frac{12}{137} \cdot x_{k-4} + h\frac{60}{137} \cdot \dot{x}_{k+1}$

Sixth-order gear 6: $x_{k+1} = \frac{360}{147}x_k - \frac{450}{147}x_{k-1} + \frac{400}{147} \cdot x_{k-2} - \frac{225}{147} \cdot x_{k-3} + \frac{72}{147} \cdot x_{k-4} - \frac{10}{147} \cdot x_{k-5} + h\frac{60}{137} \cdot \dot{x}_{k+1}$

In the case of an explicit method, after choosing the time step, we start from the initial condition $x(t_0)$ and successively compute $x(t_0 + h), x(t_0 + 2h), \cdots$ covering the entire time interval of interest. In the case of an implicit method, several iterations are made at each step. At the first iteration, an explicit method is used, and a predictor is obtained $x_{k+1}{}^{(0)} = x_k + h \cdot f(x_k)$. The value obtained for the predictor is entered on the right-hand side of the equation of the backward method obtaining a new value for $x_{k+1}{}^{(1)}$ (in the left hand side), which at the next iteration is introduced again on the right-hand side and $x_{k+1}{}^{(2)}$ is obtained and so on until $\left|x_{k+1}{}^{(N-1)} - x_{k+1}{}^{(N)}\right| < \varepsilon$ imposed. Values $x_{k+1}{}^{(1)}, x_{k+1}{}^{(2)} \cdots$ are called correctors.

In current SPICE circuit simulation programs, the forward Euler, backward Euler, trapezoidal rule, and gear method of the second order are used. In circuit design, a small simulation time is very important and integration methods of degree greater than 2 are not used because it involves many more computations and the improvement of the solution is not considerable visible.

## 5.2 Resistive (companion) models for dynamic circuit elements

The numerical methods described in the previous paragraph require writing the state equations in normal form $\dot{x} = f(x, t)$. Writing the circuit equations in this form requires the elimination of some variables and the expanding of others starting from the circuit equations. This process involves the numerical solution of some systems of linear or nonlinear algebraic equations, operation that may be affected by significantly errors and involves a certain calculation effort. This process of integrating circuit equations is simplified by replacing the dynamic elements with so-called resistive (or companion) models. By doing so, the circuit response is determined by solving a linear or nonlinear resistive circuit at each time step.

Companion models for a linear capacitor or an inductor derive from the numerical integration method used in the previous paragraph.

Starting from the relations given by the most used numerical integration method and taking into account that the state variable for the capacitor is the voltage ($i = C \cdot \frac{dv}{dt}$ or $i = C \cdot \dot{v}$ so $\dot{v} = \frac{1}{C} \cdot i$), we obtain the models in **Table 2** for the capacitor.

These equations are describing the following equivalent circuit, the capacitor companion model.

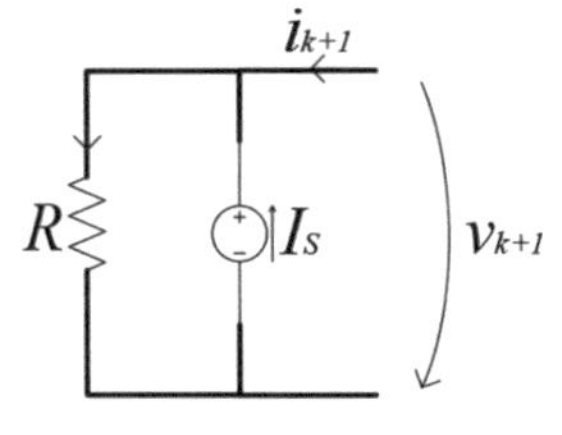

| | | | |
|---|---|---|---|
| BE | $i_{k+1} = \frac{C}{h} \cdot v_{k+1} - \frac{C}{h} \cdot v_k$ | $R = \frac{h}{C}$ | $I_S = \frac{C}{h} \cdot v_k$ |
| TR | $i_{k+1} = \frac{2 \cdot C}{h} \cdot v_{k+1} - \left(\frac{2 \cdot C}{h} \cdot v_k + i_k\right)$ | $R = \frac{h}{2 \cdot C}$ | $I_S = \frac{2 \cdot C}{h} \cdot v_k + i_k$ |
| $2^{nd}$ G | $i_{k+1} = \frac{3C}{2h} \cdot v_{k+1} - \left(\frac{9C}{8h} \cdot v_k - \frac{C}{2h} \cdot v_{k-1}\right)$ | $R = \frac{2h}{3C}$ | $I_S = \frac{9C}{8h} \cdot v_k - \frac{C}{2h} \cdot v_{k-1}$ |

**Table 2.**
*The* BE, TR *and* second G *companion models for the capacitor.*

In a similar manner but considering that the state variable for the inductor is the current ($v = L \cdot \frac{di}{dt}$ or $v = L \cdot \dot{i}$ so $\dot{i} = \frac{1}{L} \cdot v$), we obtain the models in **Table 3** for the inductor.

| | | | |
|---|---|---|---|
| BE | $i_{k+1} = \frac{h}{L} \cdot v_{k+1} + i_k$ | $R = \frac{L}{h}$ | $I_S = i_k$ |
| TR | $i_{k+1} = \frac{h}{2 \cdot L} \cdot v_{k+1} + \left(\frac{h}{2 \cdot L} \cdot v_k + i_k\right)$ | $R = \frac{2 \cdot L}{h}$ | $I_S = \frac{h}{2 \cdot L} \cdot v_k + i_k$ |
| $2^{nd}$ G | $i_{k+1} = \frac{2h}{3L} \cdot v_{k+1} + \left(\frac{3}{4} \cdot i_k - \frac{1}{3} \cdot i_{k-1}\right)$ | $R = \frac{3L}{2h}$ | $I_S = \frac{3}{4} \cdot i_k - \frac{1}{3} \cdot i_{k-1}$ |

**Table 3.**
*The* BE, TR, *and* second G *companion models for the inductor.*

These equations are describing the following equivalent circuit, the inductor companion model.

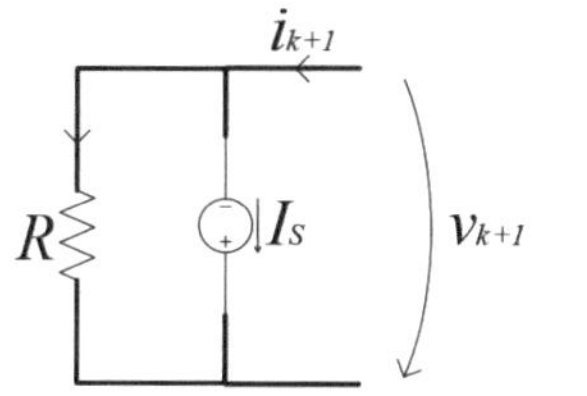


For example, using the companion models starting from the trapezoidal rule, the dynamic circuit in paragraph 2.2.1, **Figure 5**, becomes a resistive circuit but with different values at each time step for the resistors and sources in the dynamic element models (**Figure 24**).

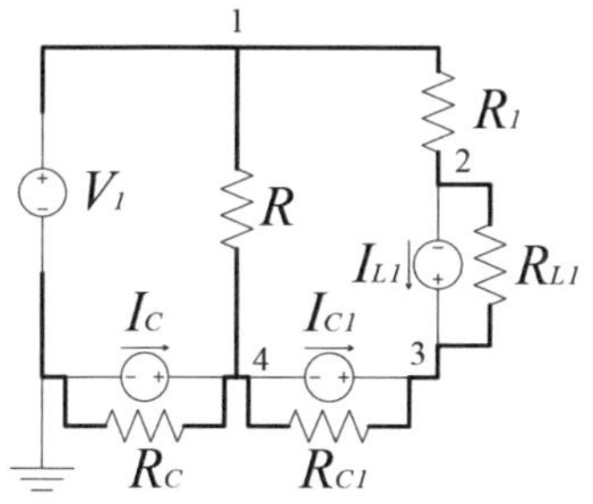


Where:
$R_{L_1} = \frac{2 \cdot L_1}{h}; I_{L_1} = \frac{h}{2 \cdot L} \cdot v_k + i_k;$
$R_{C_1} = \frac{h}{2 \cdot C_1}; I_{C_1} = \frac{2 \cdot C_1}{h} \cdot v_k + i_k;$
$R_C = \frac{h}{2 \cdot C}; I_C = \frac{2 \cdot C}{h} \cdot v_k + i_k.$

**Figure 24.**
*Dynamic elements replaced with companion models for circuit in **Figure** 5.*

The advantage of this approach is that solving a resistive circuit is much easier than a dynamic circuit, for example, using the modified nodal analysis (MNA) method [18]:

$$\begin{bmatrix} \frac{1}{R_1}+\frac{1}{R_{L_1}} & -\frac{1}{R_{L_1}} & 0 \\ -\frac{1}{R_{L_1}} & \frac{1}{R_{L_1}}+\frac{1}{R_{C_1}} & -\frac{1}{R_{C_1}} \\ 0 & -\frac{1}{R_{C_1}} & \frac{1}{R}+\frac{1}{R_{C_1}}+\frac{1}{R_C} \end{bmatrix} \cdot \begin{bmatrix} V_2 \\ V_3 \\ V_4 \end{bmatrix} = \begin{bmatrix} \frac{V_1}{R_1} - I_{L_1} \\ I_{L_1} + I_{C_1} \\ \frac{V_1}{R} + I_C - I_{C_1} \end{bmatrix}$$

This method, using a variable integration time step, is implemented in all SPICE circuit simulators.

Similar to linear dynamic element models, companion models can be built for nonlinear dynamic elements. Next, we will determine the parameters for these models for the nonlinear capacitor and the nonlinear inductor for the trapezoidal rule. Models corresponding to other numerical integration methods can be determined in a similar way.

For a nonlinear capacitor where the electric charge is a polynomial dependence of the voltage, of the form, $q = C{\cdot}(v + c_1{\cdot}v^2 + c_2{\cdot}v^3)$ the $i = \frac{dq}{dt} = \dot{q}$ the companion model can be obtained from the following.

In this case, the state variable for the capacitor is the electric charge:

$$q_{k+1} = q_k + \frac{h}{2}\cdot\left(\dot{q}_k + \dot{q}_{k+1}\right) = q_k + \frac{h}{2}\cdot\left(i_k + i_{k+1}\right)$$

$$i_{k+1} = \frac{2}{h}\cdot q_{k+1} - \frac{2}{h}\cdot q_k - i_k$$

$$i_{k+1} = \frac{2}{h}{\cdot}C{\cdot}\left(v_{k+1} + c_1{\cdot}v_{k+1}^2 + c_2{\cdot}v_{k+1}^3\right) - \left[\frac{2}{h}{\cdot}C{\cdot}\left(v_k + c_1{\cdot}v_k^2 + c_2{\cdot}v_k^3\right) + i_k\right] \tag{23}$$

This equation describes the following equivalent circuit, the nonlinear capacitor companion model.

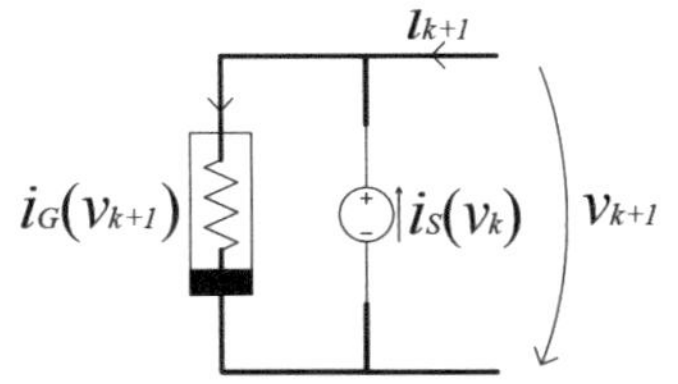


Where:
$i_G(v_{k+1}) = \frac{2}{h}{\cdot}C{\cdot}\left(v_{k+1} + c_1{\cdot}v_{k+1}^2 + c_2{\cdot}v_{k+1}^3\right)$
and
$i_S(v_k) = \frac{2}{h}{\cdot}C{\cdot}\left(v_k + c_1{\cdot}v_k^2 + c_2{\cdot}v_k^3\right) + i_k$

For a nonlinear inductor where the magnetic flux is a polynomial dependency of the current, of the form $\varphi = L{\cdot}\left(i + c_1{\cdot}i^2 + c_2{\cdot}i^3\right)$ and $v = \frac{d\varphi}{dt} = \dot{\varphi}$, the companion model can be obtained from the following.

In this case, the inductor state variable is the magnetic flux:

$$\varphi_{k+1} = \varphi_k + \frac{h}{2}\cdot\left(\dot{\varphi}_k + \dot{\varphi}_{k+1}\right) = \varphi_k + \frac{h}{2}\cdot\left(v_k + v_{k+1}\right)$$

$$v_{k+1} = \frac{2}{h}\cdot\varphi_{k+1} - \frac{2}{h}\cdot\varphi_k - v_k$$

$$v_{k+1} = \frac{2}{h}{\cdot}L{\cdot}\left(i_{k+1} + c_1{\cdot}i_{k+1}^2 + c_2{\cdot}i_{k+1}^3\right) - \left[\frac{2}{h}\cdot L{\cdot}\left(i_k + c_1{\cdot}i_k^2 + c_2{\cdot}i_k^3\right) + v_k\right] \tag{24}$$

This equation describes the following equivalent circuit, the nonlinear inductor companion model.

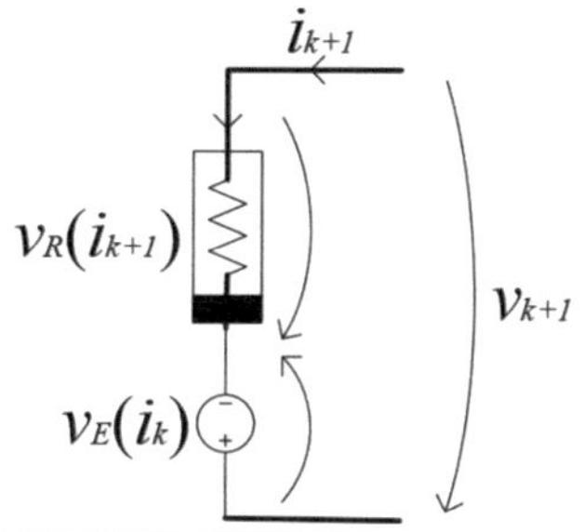


Where:
$v_R(i_{k+1}) = \frac{2}{h} \cdot L \cdot \left(i_{k+1} + c_1 \cdot i_{k+1}^2 + c_2 \cdot i_{k+1}^3\right)$
and
$v_E(i_k) = \frac{2}{h} \cdot L \cdot \left(i_k + c_1 \cdot i_k^2 + c_2 \cdot i_k^3\right) + v_k$

The resistive circuit that is solved at each time step is nonlinear in this case, even if the resistors in the circuit are linear. This circuit is solved by an iterative method, usually the Newton method.

## 6. Conclusions

All companion models contain a linear resistor with the resistance depending on the parameter of the dynamic element ($L$ or $C$) and the time step $h$, and an independent source whose parameter depends on the value of the state variable at the previous time.

Companion models of linear dynamic elements can also be built with voltage sources (the actual current source can be transformed into a voltage source). Current sources were preferred because they are suitable for the modified nodal analysis (MNA) method.

For a given time step $h$, starting from the given initial state of the dynamic elements, the circuit response is calculated at $t_0 + h$ using a *first*-order numerical integration method. In this way, the analysis of a linear dynamic circuit can be done by solving a linear resistive circuit at each time step. Starting from $t_0 + 2 \cdot h$, a *second*-order method can be used (such as the trapezoidal rule or the *second-order Gear method*). If the time step does not change, only the independent sources change in the resistive circuit, the values of the resistors remain the same. If the time step changes, the model resistance values must be recalculated.

The integration of the circuit equations is usually done with a variable time step $h$. As $h$ decreases, the resistance in the capacitor companion model decreases and the resistance in the inductor companion model increases. In the case of an $LC$ branch circuit, there are two resistors in parallel whose resistors are different by several orders of magnitude. Such a circuit cannot be solved correctly by using just simple precision. In some cases, even double-precision computations can lead to incorrect results.

## Author details

Alexandru G. Gheorghe* and Mihai E. Marin
Polytechnic University of Bucharest, Bucharest, Romania

*Address all correspondence to: alexandru.gheorghe@upb.ro

## References

[1] Chua LO, Desoer CA, Kuh ES. Linear and Nonlinear Circuits. New York: McGrawHill; 1987

[2] Chua LO, Green DN. A qualitative analysis of the behavior of dynamic nonlinear networks: Steady-state solutions of non-autonomous networks. IEEE Transactions on Circuits and Systems. 1976;**23**(9):531-550

[3] Gheorghe AG, Constantinescu F. New Topics in Simulation and Modeling of RF Circuits. Denmark: River Publishers; 2017. ISBN: 9788793379466, e-ISBN: 9788793379459

[4] Johnson DH. Origins of the equivalent circuit concept: The voltage-source equivalent. Proceedings of the IEEE. 2003;**91**(4):636-640

[5] Johnson DH. Origins of the equivalent circuit concept: The current-source equivalent. Proceedings of the IEEE. 2003;**91**(5):817-821

[6] Urbano M. Superposition theorem. In: Introductory Electrical Engineering with Math Explained in Accessible Language. US: John Wiley & Sons, Inc.; 2019. Print ISBN: 9781119580188, Online ISBN: 9781119580164. DOI: 10.1002/97811 19580164

[7] Gheorghe AG, Constantinescu F, Nitescu M. "A new algorithm for envelope following analysis", Revue Roumaine des Sciences Techniques-Serie Electrotechnique et Energetique, Nr.2011;**2**:229-236.

[8] Cristea PD, Tuduce R. State equations of circuits with excess elements - revisited. Science and Technology. 2011;**56**:219-228

[9] Marin M-E, Staicu C-S, Gheorghe AG, Constantinescu F. Generation of state equations for circuits with excess elements. In: 2020 International Symposium on Fundamentals of Electrical Engineering (ISFEE). 2020. pp. 1-4

[10] Benny Yeung. Chapter 7 Dynamic Modeling and Control of DC / DC Converters

[11] Gardner MF, Barnes JL. Transients in Linear Systems studied by the Laplace Transform. New York: Wiley; 1942

[12] Wolfram Research, Inc., Mathematica, Version 13.0.0, Champaign, IL. 2021

[13] Maple User Manual. Maplesoft, a division of Waterloo Maple Inc., 1996–2021.

[14] Gheorghe AG. Collection of Theory Problems Circuits. Politehnica Press Publishing; 2014

[15] Nagel LW, Pederson DO. SPICE (Simulation Program with Integrated Circuit Emphasis). Berkeley: University of California; 1973

[16] Star-Hspice Manual. Chapter 24: Performing Pole / Zero Analysis. Fremont, CA: Avant!, Release 1998.2.

[17] Nagel LW. SPICE2: A Computer Program to Simulate Semiconductor Circuits. Berkeley: University of California; 1975

[18] Chung-Wen H, Ruehli A, Brennan P. The modified nodal approach to network analysis. IEEE Transactions on Circuits and Systems. 1975;**22**(6):504-509. DOI: 10.1109/TCS.1975.1084079

Chapter 9

# Computation of Numerical Solution via Non-Standard Finite Difference Scheme

*Eiman Ijaz, Johar Ali, Abbas Khan, Muhammad Shafiq and Taj Munir*

## Abstract

The recent COVID-19 pandemic has brought attention to the strategies of quarantine and other governmental measures, such as lockdown, media coverage on social isolation, strengthening of public safety, etc. All these strategies are because to manage the disease as there is no vaccine and appropriate medicine for treatment. The mathematical model can assist to determine whether these intervention options are the most effective ones for illness control and how they might impact the dynamics of the disease. Motivated by this, in this manuscript, a classical order nonlinear mathematical model has been proposed to analyze the pandemic COVID-19. The model has been analyzed numerically. The suggested mathematical model is classified into susceptible, exposed, recovered, and infected classes. The non-standard finite difference scheme (NSFDS) is used to achieve the approximate results for each compartment. The graphical presentations for various compartments of the systems that correspond to some real facts are given via MATLAB.

**Keywords:** nonlinear dynamical system, COVID-19, approximate solution, NSFDS

## 1. Introduction

Many diseases have affected the human population throughout history, the most dangerous of which are viral diseases. Measles, TB, Malaria, HBV, HCV, Dengue fever, Malignant Malignancies, Spanish flu, and other diseases have resulted in millions of deaths. People have learned a memorable lesson from history. So, for controlling and reducing the rate of infections in their communities, they have established different strategies. Among the aforesaid diseases, one of the infectious diseases is COVID-19.

COVID-19 is a threatful outbreak that arose in China [1, 2] and spread throughout the globe very rapidly. It is an infectious disease caused by the virus, severe acute respiratory syndrome coronavirus 2 (SARS-CoV-2). The disease started at a seafood market in Wuhan, a big city in China, in December 2019. The disease spread in the entire city during February and March 2020. At that time, infected people were nearly 0.84 million and more than 5000 have died. Also a considerable number of infected people recovered from the said disease. The disease COVID-19 has become a pandemic due to several reasons. Some of them are (i) high transmission rate of the disease, (ii) lack of suitable vaccine and exact medicine, and (iii) the exact nature of the SARS-CoV-2 virus is still unknown. The incubation period can range from 2 to 14 days [3–12]. The majority of COVID-19 symptoms are mild, although this may increase when variants arise.

Although there were 5.94 million COVID-19 deaths that were officially reported between January 1,2020, and December 31, 2021, the excess mortality caused by the COVID-19 pandemic resulted in 18.2 million deaths globally during that time. The COVID-19 pandemic caused an excess mortality rate of 120.3 fatalities per 100,000 people worldwide. The regions of south Asia, the Middle East, north Africa, and eastern Europe had the highest number of additional deaths brought on by COVID-19. At the national level, Mexico (798, 000), Brazil (792, 000), Indonesia (736, 000), and Pakistan (664, 000) were expected to have the largest total excess mortality from COVID-19, followed by the United States (1.13 million), Russia (1.07 million), and India (4.07 million). The excess mortality rate among these nations was highest in Mexico (325.1 per 100,000) and Russia (374.6 per 100,000), and it was comparable in Brazil (186.9 per 100,000) and the USA (179.3 per 100,000).

COVID-19 symptoms differ from one person to the next. In fact, some infected people show no signs or symptoms (asymptomatic). Cough, shortness of breath or difficulty breathing, fever or chills, headaches, weariness, muscular or body aches, sore throat, loss of taste or smell, congestion or runny nose, diarrhea, and nausea or vomiting are some of the symptoms people with COVID-19 infection [9]. It's also possible that some will have additional symptoms. Many researchers, doctors, and policymakers are trying to prevent the disease from spreading. One important factor in the spreading of said disease is the migration of affected persons from one locality to another. This affects more people and hence plays a major role in the spreading. Therefore, the primary step taken by most countries is to announce city-wide lock-downs. So that some protective measures should be taken to minimize the greatest possible loss of human lives [13]. On an international level, banned air traffic for an unknown period of time. Keeping in mind that in the past such outbreak not only led to the greatest loss of human lives but also damaged the economy very badly throughout the world. Therefore, scientists and researchers are trying their best to put their part in the investigation of a cure for the COVID-19 outbreak. It is clear from a medical engineering point of view that infectious diseases can be better understood by using the mathematical model. In the last many decades, mathematical modeling is one of the important areas of research [14–55]. To understand the dynamics of COVID-19, it is essential to formulate mathematical models that can assist in the estimation of the transmissibility and dynamic of the virus transmission. Also, the majority of real-world problems, such as infectious diseases, are nonlinear in nature. As a result, nonlinear mathematical models that describe a variety of real-world issues have piqued interest for decades. In this regard, various models were formulated or updated. Also, several types of research focusing on mathematical modeling of COVID-19 have been considered recently. Some models that have recently been considered in this regard are [56–59]. Motivated by the above work, we are going to investigate the COVID-19 mathematical model (see 4) numerically under NSFDS.

## 2. Preliminaries

In numerical analysis, NSFDS is a general set of methods that gives numerical solutions to differential equations by discretizing the data. Many real-life problems are modeled by differential equations, for which analytical solutions are difficult to find out efficiently. Several researchers have tried different ways (e.g., via Finite Element Methods, Standard Finite Difference Methods, Spline Approximation Methods, etc). Nowadays, NSFDS is playing an important role in solving the real-life problems governed by ODEs and/or by PDEs. In science and engineering, many differential models for which the existing methodologies do not give reliable results, NSFDS are solving them competitively.

Here we derive the suggested scheme for simple problems as let

$$\frac{dy}{dt} = f(t, y) \tag{1}$$

then NSFD equation is

$$\frac{y_{k+1} - y_k}{h} = f(t, y(k)),$$
$$y_{k+1} = y_k + hf(t, y(k)).$$

**Definition 1.** *A successful example of a NSFD equation is one setup for a combustion model*

$$\frac{dw}{dt} = w^2(1 - w). \tag{2}$$

The NSFD equation would be

$$\frac{w_{k+1} - w_k}{h} = w_k^2 - w_k^3. \tag{3}$$

## 3. Formulation of proposed model

A model is formulated that further divides the entire population into different classes given as:

individuals who have high chance of getting an infection are placed in susceptible class $S$, individuals who are in close contact with COVID-19 environment are placed in exposed class $E$, individuals having the symptoms of COVID-19 are placed in infected class $I$ and $R$ recovered class includes recovered individuals. A mathematical model of COVID-19 is described by the following system of differential eqs. [30].

$$\begin{cases} \frac{d}{dt}S(t) = \gamma - k(1 + \alpha I(t))S(t)I(t) - \varepsilon S(t), \\ \frac{d}{dt}E(t) = k(1 + \alpha I(t))S(t)I(t) - (\varepsilon + \delta)E(t), \\ \frac{d}{dt}I(t) = \eta + \delta E(t) - (\upsilon + \varepsilon + \beta)I(t), \\ \frac{d}{dt}R(t) = \beta I(t) - \varepsilon R(t). \end{cases} \tag{4}$$

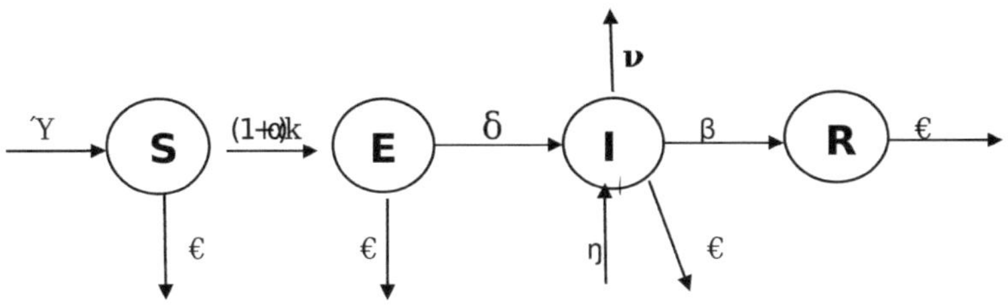


**Figure 1.**
*A flow chart of the proposed model.*

With initial conditions given by

$$S(0) = S_0,\ E(0) = E_0,\ I(0) = I_0,\ R(0) = R_0.$$

The description of above model is given in **Figure 1**.

## 4. Algorithm for approximate solution of the considered model

To compute the required approximate solution, using general form of NSFD on (4), we have

$$\begin{cases} \dfrac{S_{n+1}(t) - S_n(t)}{h} = \gamma - k(1 + \alpha I_n(t))S_n(t)I_n(t) - \varepsilon S_n(t), \\ \dfrac{E_{n+1}(t) - E_n(t)}{h} = k(1 + \alpha I_n(t))S_n(t)I_n(t) - (\varepsilon + \delta)E_n(t), \\ \dfrac{I_{n+1}(t) - I_n(t)}{h} = \eta + \delta E_n(t) - (\upsilon + \varepsilon + \beta)I_n(t), \\ \dfrac{R_{n+1}(t) - R_n(t)}{h} = \beta I_n(t) - \varepsilon R_n(t). \end{cases} \tag{5}$$

$$\begin{cases} S_{n+1}(t) = S_n(t) + h(\gamma - k(1 + \alpha I_n(t))S_n(t)I_n(t) - \varepsilon S_n(t)), \\ E_{n+1}(t) = E_n(t) + h(k(1 + \alpha I_n(t))S_n(t)I_n(t) - (\varepsilon + \delta)E_n(t)), \\ I_{n+1}(t) = I_n(t) + h(\eta + \delta E_n(t) - (\upsilon + \varepsilon + \beta)I_n(t)), \\ R_{n+1}(t) = R_n(t) + h(\beta I_n(t) - \varepsilon R_n(t)). \end{cases} \tag{6}$$

Now putting n = 0, 1, 2 …. in (6), we get few terms of the approximate solution as

$$\begin{cases} S_1(t) = S_0(t) + h(\gamma - k(1 + \alpha I_0(t))S_0(t)I_0(t) - \varepsilon S_0(t)), \\ E_1(t) = E_0(t) + h(k(1 + \alpha I_0(t))S_0(t)I_0(t) - (\varepsilon + \delta)E_0(t)), \\ I_1(t) = I_0(t) + h(\eta + \delta E_0(t) - (\upsilon + \varepsilon + \beta)I_0(t)), \\ R_1(t) = R_0(t) + h(\beta I_0(t) - \varepsilon R_0(t)). \end{cases} \tag{7}$$

$$\begin{cases} S_2(t) = S_1(t) + h(\gamma - k(1 + \alpha I_1(t))S_1(t)I_1(t) - \varepsilon S_1(t)), \\ E_2(t) = E_1(t) + h(k(1 + \alpha I_1(t))S_1(t)I_1(t) - (\varepsilon + \delta)E_1(t)), \\ I_2(t) = I_1(t) + h(\eta + \delta E_1(t) - (\upsilon + \varepsilon + \beta)I_1(t)), \\ R_2(t) = R_1(t) + h(\beta I_1(t) - \varepsilon R_1(t)). \end{cases} \quad (8)$$

$$\begin{cases} S_3(t) = S_2(t) + h(\gamma - k(1 + \alpha I_2(t))S_2(t)I_2(t) - \varepsilon S_2(t)), \\ E_3(t) = E_2(t) + h(k(1 + \alpha I_2(t))S_2(t)I_2(t) - (\varepsilon + \delta)E_2(t)), \\ I_3(t) = I_2(t) + h(\eta + \delta E_2(t) - (\upsilon + \varepsilon + \beta)I_2(t)), \\ R_3(t) = R_2(t) + h(\beta I_2(t) - \varepsilon R_2(t)). \end{cases} \quad (9)$$

and so on. Similarly, the other terms may be computed.

## 5. Numerical interpretation

To present the concerned approximate solutions computed above of the model under consideration, we use numerical values for the parameters in given in **Table 1**. Based on reported data, the initial condition is set as [45]

$$(\mathbb{S}(0), \mathbb{E}(0), \mathbb{I}(0), \mathbb{R}(0)) = (32.37 million,\ 12 million, 0.001523 million, 0.005025 million).$$

After putting the numerical values in Eq. (6), we obtained the following results.
**Case (1) n = 0**

$$\begin{cases} \mathbb{S}_1(t) = 3.2018 \times 10^7, \\ \mathbb{E}_1(t) = 1.3738 \times 10^6, \\ \mathbb{I}_1(t) = 4.5718 \times 10^3, \\ \mathbb{R}_1(t) = 5.0163 \times 10^3. \end{cases} \quad (10)$$

| Parameters | Description of parameters | Numerical value |
|---|---|---|
| $\gamma$ | Tested negative population | $0.250281 \times 10^{-6}$ |
| $\eta$ | Tested positive population | $0.006656 \times 10^{-6}$ |
| $k$ | The infection rate | 0.000024 |
| $\alpha$ | Rate of individual lose immunity | 0.01182 |
| $\varepsilon$ | Natural death rate | $0.0000004 \times 10^{-6}$ |
| $\nu$ | Death rate due to C0VID-19 | 0.016 |
| $\delta$ | Infected rate | 0.025 |
| $\beta$ | Recovered rate | 0.75 |

**Table 1.**
*Numerical values of parameters.*

And similarly from Eqs. (8) and (9), we get
**Case (2) n = 1**

$$\begin{cases} S_2(t) = 2.9958 \times 10^7, \\ E_2(t) = 3.3015 \times 10^6, \\ I_2(t) = 4.9285 \times 10^3, \\ R_2(t) = 5.0305 \times 10^3. \end{cases} \tag{11}$$

**Case (3) n = 2**

$$\begin{cases} S_3(t) = 2.7741 \times 10^7, \\ E_3(t) = 5.3871 \times 10^6, \\ I_3(t) = 5.7629 \times 10^3, \\ R_3(t) = 5.0473 \times 10^3. \end{cases} \tag{12}$$

**Case (4) n = 3**

$$\begin{cases} S_4(t) = 2.4980 \times 10^7, \\ E_4(t) = 8.0161 \times 10^6, \\ I_4(t) = 7.1091 \times 10^3, \\ R_4(t) = 5.0703 \times 10^3. \end{cases} \tag{13}$$

**Case (5) n = 4**

$$\begin{cases} S_5(t) = 2.1258 \times 10^7, \\ E_5(t) = 1.1606 \times 10^7, \\ I_5(t) = 9.0967 \times 10^3, \\ R_5(t) = 5.1033 \times 10^3. \end{cases} \tag{14}$$

In **Figures 2–5**, we have provided a graphical representation of different classes for the proposed model. We concluded that by taking a few terms of the series solutions we can efficiently describe the proposed model. We see in the figures that the

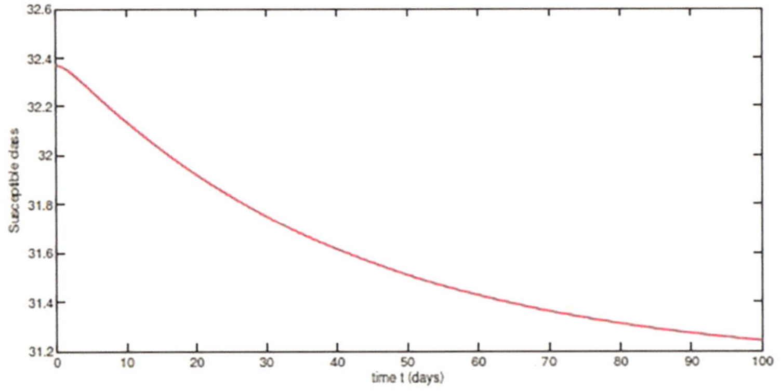


**Figure 2.**
*Dynamics of susceptible class.*

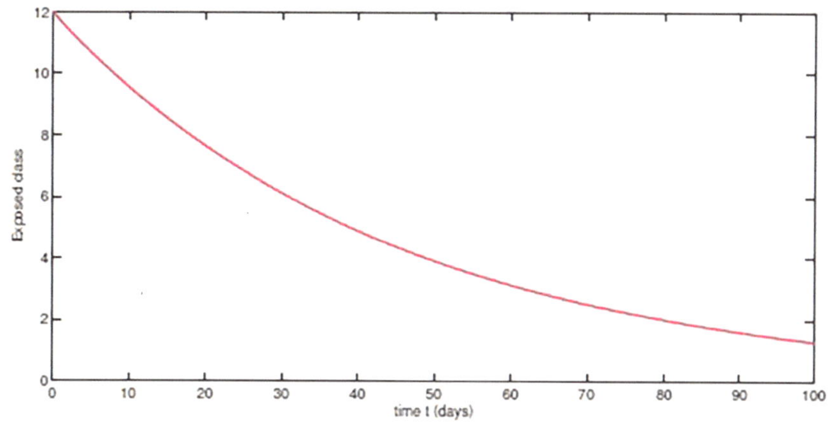


**Figure 3.**
*Dynamics of exposed class.*

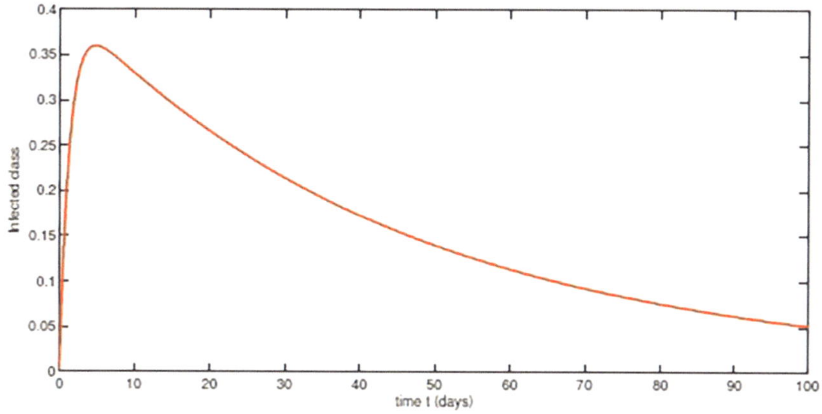


**Figure 4.**
*Dynamics of infected class.*

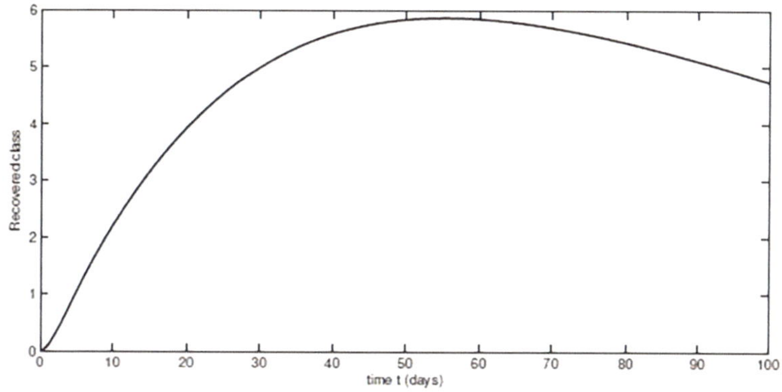


**Figure 5.**
*Dynamics of recovered class.*

susceptible class is decreasing as a result increase in infection occurred but due to vaccination and other precautions there occurred an increase in the recovered class. Further, we compare our results with the usual RK4 method numerical results for the given data in **Table 1** in **Figures 6–9** respectively. We see that the solution through the NSFDS and RK4 method agrees very well.

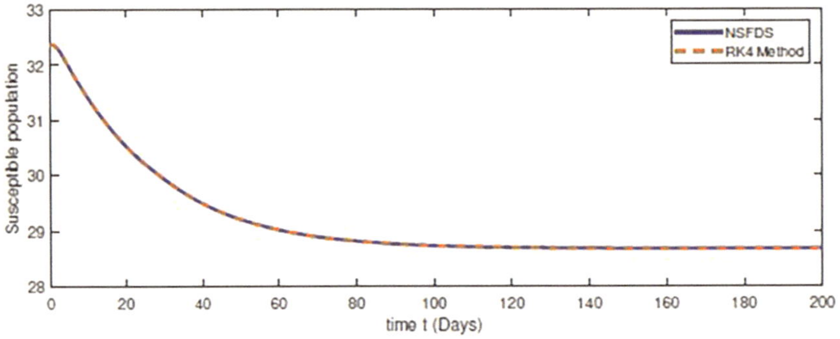


**Figure 6.**
*Comparison of the approximate solution for the susceptible class at NSFS and RK4.*

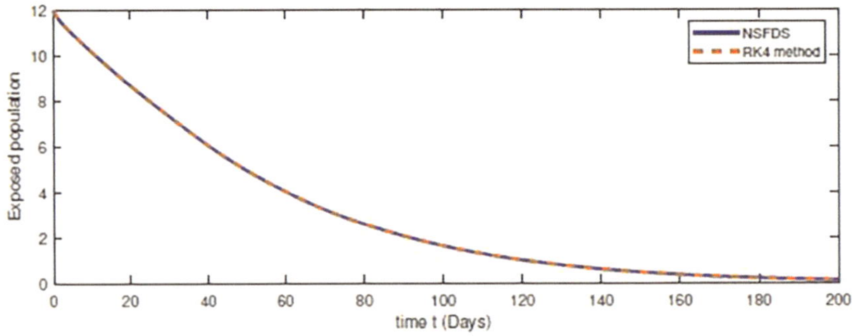


**Figure 7.**
*Comparison of the approximate solution for the exposed class at NSFS and RK4.*

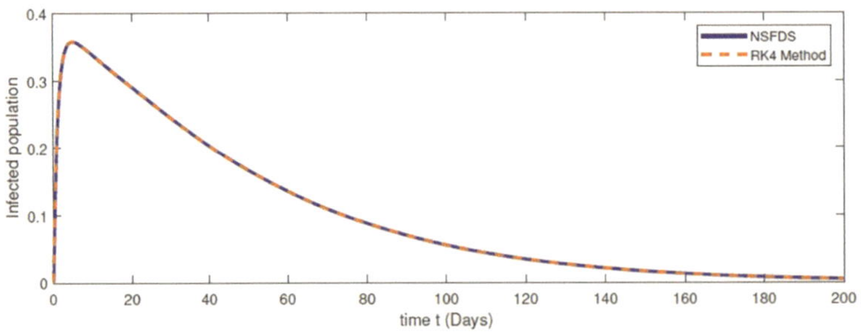


**Figure 8.**
*Comparison of the approximate solution for the infected class at NSFS and RK4.*

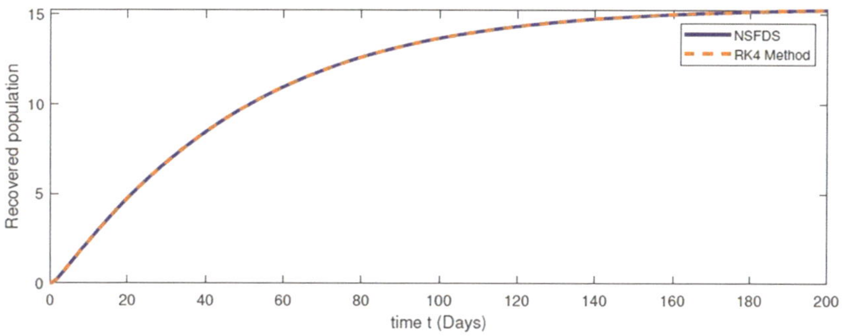


**Figure 9.**
*Comparison of the approximate solution for the recovered class at NSFS and RK4.*

## 6. Some explanation and concluding remarks

In this work, we have studied a four-compartmental mathematical model based on a system of ordinary differential equations to study the dynamics of COVID-19 through the NSFDS method. With the help of the said technique, we develop an algorithm to discretize the data to find an approximate solution to the proposed problem. Using some real values for the parameters and initial data, we compute a few terms and approximate solutions corresponding to a different compartment. We plot our approximate solutions for different compartments graphically using MATLAB. We concluded that by taking a few terms of the solutions, we can efficiently describe the proposed model. As compared to RK4 and Euler methods, NSFD method is easy to implement. The computational cost is low and also good for time-saving in the future, one can extend the current study for mathematical models under nonsingular type derivatives. Finally, we have given a comparison between the approximate solution at NSFD method and RK4 method. We see that both solutions agreed very well.

## Author details

Eiman Ijaz[1], Johar Ali[1], Abbas Khan[1], Muhammad Shafiq[1] and Taj Munir[2*]

1 Department of Mathematics, University of Malakand, Pakistan

2 Abdus Salam School of Mathematical Sciences G.C. University Lahore Punjab, Pakistan

*Address all correspondence to: ehuzaifa@gmail.com; taj_math@hotmail.com

## References

[1] Chan JF-W et al. Genomic characterization of the 2019 novel human-pathogenic coronavirus isolated from a patient with atypical pneumonia after visiting Wuhan. Emerging Microbes & Infections. 2020;**9**(1): 221-236

[2] World Health Organization. Coronavirus disease 2019 (COVID-19) Situation Report-62. 2019

[3] Riou J, Althaus CL. Pattern of early human-to-human transmission of Wuhan 2019 novel coronavirus (2019-nCoV), December 2019 to January 2020. Eurosurveillance. 2020;**25**(4):2000058

[4] Hurwitz JL. Viruses and the sars-cov-2/covid-19 pandemic of 2020. Viral Immunology. 2020;**33**(4):251-252

[5] Ge XY et al. Isolation and characterization of a bat SARS–like coronavirus that uses the ACE2 receptor. Nature. 2013;**503**:535-538

[6] Zhou P, Yang X-L, Wang X-G, Ben H, Zhang L, Zhang W, et al. A pneumonia outbreak associated with a new coronavirus of probable bat origin. Nature. 2020;**579**(7798):270-273

[7] Sha H, Sanyi T, Libin R. A discrete stochastic model of the covid-19 outbreak, Forecast and control. Mathematical Bioscience Engineering. 2020;**17**(4):2792-2804

[8] Fisher D, Heymann D. The novel coronavirus outbreak causing covid-19. BMC Medicine. 2020;**18**(1):1-3

[9] Forida P et al. The symptoms, contagious process, prevention and post treatment of Covid-19. European Journal of Physiotherapy and Rehabilitation Studies. 2020;**2020**:11

[10] World Health Organization. Advice on the use of masks in the context of COVID-19: Interim guidance. 2020

[11] McAloon C et al. Incubation period of COVID-19, a rapid systematic review and meta-analysis of observational research. BMJ Open. 2020;**10**(8):e039652

[12] Quesada JA et al. Incubation period of COVID-19, a systematic review and meta-analysis. Revista Clinica Espanola (English Edition). 2021;**221**(2):109-117

[13] Lin Q et al. (COVID-19) outbreak in Wuhan, China with individual reaction and governmental action. International Journal of Infectious Diseases. 2019; **93**(2020):211-216

[14] Li Q et al. Early transmission dynamics in Wuhan, China, of novel coronavirus-infected pneumonia. New England Journal of Medicine. 2020;**382**: 1199-1207

[15] Alqudah M, Abdeljawad T, Eiman Q, Madlal K, Shah FJ. Existence theory and approximate solution to prey-predator coupled system involving non singular kernel type derivative. Advanced in Difference Equation. 2020;**1**:1-10

[16] Moaddy K, Momani S, Hashim I. The non-standard finite difference scheme for linear fractional PDEs in fluid mechanics. Computers & Mathematics with Applications. 2011;**61**(4):1209-1216

[17] Mickens RE. Applications of nonstandard finite difference schemes. Singapore: World Scientific; 2000

[18] Adekanye O, Washington T. Nonstandard finite difference scheme for a Tacoma Narrows Bridge model. Applied Mathematical Modelling. 2018;**62**:223-236

[19] Korpusik A. A nonstandard finite difference scheme for a basic model of cellular immune response to viral infection. Communications in Nonlinear Science and Numerical Simulation. 2017; **43**:369-384

[20] Mickens RE. A nonstandard finite difference scheme for a Fisher PDE having nonlinear diffusion. Computers and Mathematics with Applications. 2003;**45**:429-436

[21] Hajipour M, Jajarmi A, Baleanu D. An efficient nonstandard finite difference scheme for a class of fractional chaotic systems. Journal of Computational and Nonlinear Dynamics. 2018;**13**(2)

[22] Xu J, Geng Y, Hou J. A non-standard finite difference scheme for a delayed and diffusive viral infection model with general nonlinear incidence rate. Computers and Mathematics with Applications. 2017;**74**(8):1782-1798

[23] Qin W, Wang L, Ding X. A nonstandard finite difference method for a hepatitis B virus infection model with spatial diffusion. Journal of Difference Equations and Applications. 2014; **20**(12):1641-1651

[24] Manna K. A nonstandard finite difference scheme for a diffusive HBV infection model with capsids and time delay. Journal of Difference Equations and Applications. 2017;**23**(11):1901-1911

[25] Manna K, Chakrabarty SP. Global stability and a nonstandard finite difference scheme for a diffusion driven HBV model with capsids. Journal of Difference Equations and Applications. 2015;**21**(10):918-933

[26] Elsheikh S, Ouifki R, Patidar KC. A nonstandard finite difference method to solve a model of HIV–Malaria co–infection. Journal of Difference Equations and Applications. 2014;**20**(3): 354-378

[27] Tadmon C, Foko S. Nonstandard finite difference method applied to an initial boundary value problem describing hepatitis B virus infection. Journal of Difference Equations and Applications. 2020;**26**(1):122-139

[28] Bisheh-Niasar M, Arab Ameri M. Moving meshnonstandard finite difference method for non–linear heat transfer in a thin finite rod. Journal of Applied and Computational Mechanics. 2018;**4**(3):161-166

[29] Zafar ZU, Abadin NA, Younas S, Abdelwahab SF, Nisar KS. Numerical investigations of stochastic HIV/AIDS infection model. Alexandria Engineering Journal. 2021;**60**(6):5341-5363

[30] Yang Y, Zhou J, Ma X, Zhang T. Nonstandard finite difference scheme for a diffusive within–host virus dynamics model with both virus–to–cell and cell–to–cell transmissions. Computers Mathematics with Applications. 2016;**72**(4):1013-1020

[31] Singh H. Analysis for fractional dynamics of Ebola virus model. Chaos Solitons & Fractals. 2020;**138**: 109992

[32] Singh H, Singh CS. A reliable method based on second kind Chebyshev polynomial for the fractional model of Bloch equation. Alexandria Engineering Journal. 2018;**57**(3):1425-1432

[33] Singh H. Operational matrix approach for approximate solution of fractional model of Bloch equation. Journal of King Saud University–Science. 2017;**29**(2):23-240

[34] Singh H, Pandey R, Srivastava H. Solving non-linear fractional variational

problems using jacobi polynomials. Mathematics. 2019;**7**(3):224

[35] Singh H, Srivastava HM. Numerical investigation of the fractional order liénard and duffing equations arising in oscillating circuit theory. Frontier in Physics. 2020;**8**:120

[36] Singh H, Sahoo MR, Singh OP. Numerical method based on Galerkin approximation for the fractional advection–dispersion equation. International Journal of Applied and Computational Mathematics. 2017;**3**(3): 2171-2187

[37] Zhang Y. Initial boundary value problem for fractal heat equation in the semi-infinite region by Yang–Laplace transform. Thermal Science. 2014;**18**(2): 677-681

[38] Miller KS, Ross B. An Introduction to the Fractional Calculus and Fractional Differential Equations. New York: Wiley; 1993

[39] Eltayeb H, Kiliçman A. A note on solutions of wave, Laplace's and heat equations with convolution terms by using a double Laplace transform. Applied Mathematics Letters. 2008; **21**(12):1324-1329

[40] Spiga G, Spiga M. Two-dimensional transient solutions for crossflow heat exchangers with neither gas mixed. Journal of Heat Transfer-transactions of the ASME. 1987;**109**(2):281-286

[41] Khan T, Shah K, Khan RA, Khan A. Solution of fractional order heat equation via triple Laplace transform in 2 dimensions. Mathematical Methods in the Applied Sciences. 2018;**4**(2):818-825

[42] Shah K, Khalil H, Khan RA. Analytical solutions of fractional order diffusion equations by natural transform method. Iranian Journal of Science and Technology, Transactions A: Science. 2018;**42**(3):1479-1490

[43] Singh H, Ghassabzadeh FA, Tohidi E, Cattani C. Legendre spectral method for the fractional Bratu problem. Mathematical Methods in the Applied Sciences. 2020;**43**(9):5941-5952

[44] Singh H, Srivastava HM. Jacobi collocation method for the approximate solution of some fractional order Riccati differential equations with variable coefficients. Physica A. 2019;**523**: 1130-1149

[45] Singh H, Srivastava HM, Kumar D. A reliable algorithm for the approximate solution of the nonlinear Lane–Emden type equations arising in astrophysics. Numerical Methods for Partial Differential Equations. 2018;**34**(5): 1524-1555

[46] Singh J, Jassim HK, Kumar D. An efficient computational technique for local fractional Fokker Planck equation. Physica A. 2020;**555**(1):124525

[47] Ahmad B, Sivasundaram S. On four–point nonlocal boundary value problems of nonlinear integro–differential equations of fractional order. Applied Mathematics and Computation. 2010; **217**:480-487

[48] Bai Z. On positive solutions of a nonlocal fractional boundary value problem. Nonlinear Analysis. 2010;**72**: 916-924

[49] Khan RA, Shah K. Existence and uniqueness of solutions to fractional order multi-point boundary value problems. Communications in Applied Analysis. 2015;**19**:515-526

[50] Shah K, Ali N, Khan RA. Existence of positive solution to a class of

fractional differential equations with three point boundary conditions. Mathematics Science Letter. 2016;**5**(3): 291-296

[51] Wang J, Zhou Y, Wei W. Study in fractional differential equations by means of topological degree methods. Numerical Functional Analysis Optimum. 2012;**33**:216-238

[52] Brauer F, Castillo-Chavez C. Mathematical Models in Population Biology and Epidemiology. New York: Springer; 2001

[53] Hethcote HW. The mathematics of infectious diseases. SIAM Review. 2000; **42**:599

[54] Hethcote HW, Van Ark JW. Modeling HIV Transmission and AIDS in the United States. Berlin, Heidelberg, New York: Springer; 1992

[55] Lu H, Stratton CW, Tang YW. Outbreak of pneumonia of unknown Etiology in Wuhan China: The mystery and the miracle. Journal of Medical Virology. 2020;**2020**:1234-1260

[56] Lin Q et al. A conceptual model for the coronavirus disease 2019 (COVID-19) outbreak in Wuhan, China with individual reaction and governmental action. International Journal of Infectious Diseases. 2020;**93**:211-216

[57] Yousaf M et al. Statistical analysis of forecasting COVID-19 for upcoming month in Pakistan. Chaos, Solitons & Fractals. 2020;**2020**:109926

[58] Shah K et al. Qualitative analysis of a mathematical model in the time of COVID-19. BioMed Research International. 2020;**2020**:11

[59] Abdo MS et al. On a comprehensive model of the novel coronavirus (COVID-19) under Mittag-Leffler derivative. Chaos, Solitons & Fractals. 2020;**2020**:109867